Lecture Notes in Mathematics

A collection of informal reports and seminars

Edited by A. Dold, Heidelberg and B. Eckmann, Zürich

279

R. Bott

S. Gitler

I. M. James

Lectures on Algebraic and Differential Topology

Delivered at the II. ELAM

Springer-Verlag

Berlin · Heidelberg · New York 1972

R. Bott
Harvard University, Cambridge, MA/USA

S. Gitler
Centro de Investigación, Mexico City, Mexico

I. M. James
Mathematical Institute, Oxford/England

AMS Subject Classifications (1970): 14F05, 14F10, 58A30, 57D30, 57D35, 57D40, 55F25, 55B20, 55F99, 55G10, 55G20, 55G45

ISBN 3-540-05944-X Springer-Verlag Berlin · Heidelberg · New York
ISBN 0-387-05944-X Springer-Verlag New York · Heidelberg · Berlin

Offsetdruck: Julius Beltz, Hemsbach/Bergstr.

PREFACE

The Second Latin American School in Mathematics was held during the month of July at the Centro de Investigación del IPN, in Mexico City and was dedicated to the memory of Heinz Hopf. The School was held under the auspices of the Centro de Investigación del IPN, Organization of American States, The National Science Foundation and the Consejo Nacional de Ciencia y Tecnología.

These notes contain three of the seven series of lectures given at the School. We hope that the others will appear at some future date. We thank the participants and the lecturers for having made the School a most stimulating mathematical event.

Mexico, D.F., April 14, 1972.

Samuel Gitler
Editor

CONTENTS

LECTURES ON CHARACTERISTIC CLASSES AND FOLIATIONS

by Raoul Bott

(Notes by Lawrence Conlon)

1. Some basic notions. In this section we presume some familiarity with vector bundles and differential forms. Brief discussions of these two notions will be found in §2 and §3 respectively. We also presume (both here and throughout the notes) some familiarity with differentiable manifolds.

Let M and N be smooth manifolds of respective dimensions m and n. Let $f : M \to N$ be a smooth map. For each $x \in M$ one has the linear map

$$df_x : T_x(M) \to T_{f(x)}(N) .$$

(1.1) Definition. f as above is called an <u>immersion</u> if df_x is injective, $\forall\, x \in M$. f is called a <u>submersion</u> if df_x is surjective, $\forall\, x \in M$.

Of these two notions we are here primarily interested in submersions. By the implicit function theorem, df_x surjective implies that $f^{-1}(f(x))$ is a smooth submanifold of M of dimension $m - n$. Thus, a submersion $f : M \to N$ decomposes M into $(m - n)$-dimensional "fibers" $f^{-1}(y)$, $y \in N$. This is an example of the sort of structure we will call a foliation. The crucial property of this example is the fact that locally one can choose coordinates on M and N such that f becomes the canonical projection

$$R^n \times R^{m-n} \to R^n .$$

We further notice that the vector spaces $\operatorname{Ker}(df_x)$ unite to form a smooth subbundle $\operatorname{Ker}(df) \subset T(M)$ (equivalently, a smooth $(m - n)$-dimensional distribution on M in the sense of [1, p. 10]).

(1.2) **Definition.** A smooth subbundle $E \subset T(M)$ is called integrable iff locally E is the kernel of the differential of a submersion. An integrable subbundle of $T(M)$ is also called a foliation of M.

Notice that a foliation of M gives rise to an open cover $\{U_\alpha\}$ and submersions $\{f_\alpha : U_\alpha \to R^q\}$ such that

$$\mathrm{Ker}(df_\alpha)|U_\alpha \cap U_\beta = \mathrm{Ker}(df_\beta)|U_\alpha \cap U_\beta$$

for all α,β. Furthermore, if $x \in U_\alpha \cap U_\beta$ it can be shown that there is an open neighborhood $W \subset U_\alpha \cap U_\beta$ of x such that $f_\alpha^{-1}(f_\alpha(x)) \cap W = f_\beta^{-1}(f_\beta(x)) \cap W$. Conversely, such a structure clearly determines a unique foliation of M.

We show by an example that not every subbundle $E \subset T(M)$ is integrable. Let $M = R^3 - \{0\}$ and consider the 1-form $\varphi = x\cdot dy + y\cdot dz + z\cdot dx$. φ is nowhere zero on M, hence $E = \mathrm{Ker}(\varphi)$ is a two dimensional smooth subbundle of $T(M)$. If E were integrable we could find $\{U_\alpha, f_\alpha : U_\alpha \to R^1\}$ as above such that

$$\mathrm{Ker}(df_\alpha) = \mathrm{Ker}(\varphi)|U_\alpha, \quad \forall\alpha .$$

Thus $\varphi|U_\alpha = h\cdot df_\alpha$ where $h : U_\alpha \to R$ is C^∞ and nowhere zero. On U_α this gives

$$d\varphi = dh \wedge df_\alpha = \frac{dh}{h} \wedge hdf_\alpha = \eta \wedge \varphi$$

where η is a 1-form. Thus, on each U_α

$$\varphi \wedge d\varphi = \varphi \wedge \eta \wedge \varphi = 0$$

hence $\varphi \wedge d\varphi = 0$ identically on M. But direct computation shows

$$\varphi \wedge d\varphi = (x+y+z)dx \wedge dy \wedge dz$$

contradicting the above. E, therefore, cannot be integrable.

(1.3) Theorem. (Deahna, Clebsch, Frobenius). Let $E \subset T(M)$ be a smooth subbundle described locally as the simultaneous kernel of everywhere linearly independent 1-forms $\theta_1,\ldots,\theta_q$ (i.e., for x in the domain of the θ_i's, $E_x = \{v \in T_x(M) : \theta_1(v) = \cdots = \theta_q(v) = 0\}$) . Then E is integrable iff each $d\theta_i$ belongs to the ideal generated by $\theta_1,\ldots,\theta_q$.

An equivalent formulation of this theorem is the following.

(1.3)′ Theorem. A smooth subbundle $E \subset T(M)$ is integrable iff for any two smooth sections X and Y of E, the vector field $[X,Y]$ is also a section of E.

Remark that a simple corollary of either formulation, which also follows immediately from the standard existence and uniqueness theorem for ordinary differential equations, is that any smooth 1-dimensional subbundle $E \subset T(M)$ is integrable.

Theorems (1.3) and (1.3)′, usually referred to jointly as the "Frobenius Theorem", give a complete and satisfying answer to the (essentially local) question of integrability of a subbundle $E \subset T(M)$. We next modify the integrability question to obtain an essentially global problem to which no complete answer has been given.

(1.4) If $E \subset T(M)$ is a subbundle, does there exist an integrable subbundle $E' \subset T(M)$ such that E' and E are isomorphic as vector bundles?

One of the main aims of these lectures will be to give topological obstructions to the existence of E' in terms of the characteristic classes of E.

<u>References</u>

1. S. Kobayashi and K. Nomizu, <u>Foundations of Differential Geometry</u>, Vol. I, 1963, Interscience Publishers, New York.

2. Vector bundles and Pontryagin classes

While a certain basic knowledge of vector bundles might well be presumed in these notes, we wish here to review the concept from a point of view particularly suited to our purposes.

(2.1) **Definition.** An n-dimensional (real) vector bundle over a space X is a continuous map $\pi : E \to X$ such that

1) $\exists$ open cover $\{U_\alpha\}_{\alpha \in A}$ of X and homeomorphisms $\varphi_\alpha : U_\alpha \times R^n \to \pi^{-1}(U_\alpha)$, $\forall \alpha \in A$.

2) The diagram (where p is the standard projection)

$$\begin{array}{ccc} U_\alpha \times R^n & \xrightarrow{\varphi_\alpha} & \pi^{-1}(U_\alpha) \\ & {\scriptstyle p}\searrow \quad \swarrow{\scriptstyle \pi} & \\ & U_\alpha & \end{array}$$

is commutative, $\forall \alpha \in A$.

3) For any $\alpha, \beta \in A$ with $U_\alpha \cap U_\beta \neq \emptyset$ the map

$$\varphi_\beta^{-1} \circ \varphi_\alpha : (U_\alpha \cap U_\beta) \times R^n \to (U_\alpha \cap U_\beta) \times R^n$$

is of the form

$$\varphi_\beta^{-1} \circ \varphi_\alpha(x,v) = (x, g_{\beta\alpha}(x) \cdot v)$$

where $g_{\beta\alpha} : U_\alpha \cap U_\beta \to GL_n = GL(n,R)$ is continuous. Here, of course, $GL(n,R)$ denotes the group of $n \times n$ nonsingular matrices.

By a certain abuse, we often call E itself the vector bundle over X. More properly E is called the total space of the bundle.

Note that each fiber $\pi^{-1}(x)$, $x \in X$, has canonically the structure of an n-dimensional vector space over R, although $\pi^{-1}(x)$ does not usually have a canonical coordinatization as R^n. As a general principle, operations on vector spaces extend to vector bundles. Thus we may form the direct sum $E \oplus E'$ of bundles over X, the quotient E/E' of a bundle by a subbundle, etc., obtaining new bundles over X.

Remark that the maps $g_{\alpha\beta}$ satisfy the following "cocycle condition".

(2.2) On $U_\alpha \cap U_\beta \cap U_\gamma$, $g_{\alpha\beta} \cdot g_{\beta\gamma} = g_{\alpha\gamma}$.

Conversely, suppose an open cover $\{U_\alpha\}_{\alpha \in A}$ of X is given together with continuous maps

$$g_{\alpha\beta} : U_\alpha \cap U_\beta \to GL_n$$

defined for all $\alpha, \beta \in A$ such that $U_\alpha \cap U_\beta \neq \emptyset$ and, further, satisfying (2.2). On the disjoint union $\coprod_{\alpha \in A} U_\alpha \times R^n$ make the identifications

$$(x,v) \sim (x, g_{\beta\alpha}(x)v)$$

$\forall\, x \in U_\alpha \cap U_\beta \neq \emptyset$, $\forall\, v \in R^n$. The resulting quotient space F and the natural map $\rho : F \to X$ define a vector bundle over X.

(2.3) **Definition.** If $\pi : E \to X$ and $\rho : F \to X$ are vector bundles, a continuous map $f : E \to F$ is said to be linear if

1) $$\begin{array}{ccc} E & \xrightarrow{f} & F \\ & \pi\searrow \quad \swarrow\rho & \\ & X & \end{array}$$ commutes

2) f is linear on each fiber.

Furthermore, we will say that E and F are isomorphic (and write $E \simeq F$) iff there are linear maps $f : E \to F$ and $g : F \to E$ such that both $f \circ g$ and $g \circ f$ are identity maps.

(2.4) **Exercise.** Let $\pi : E \to X$ be an n-dimensional vector bundle over X. Using the definition (2.1) take an associated cocycle $\{g_{\alpha\beta}\}$ and construct a new bundle $\rho : F \to X$ as above and show that $E \simeq F$.

Let $\mathcal{H}$ be a separable real Hilbert space. Define BGL_n to be the set of n-dimensional subspaces of $\mathcal{H}$ and topologize this set in some "reasonable" way (e.g., define a metric on BGL_n by taking the distance between two n-dimensional subspaces of $\mathcal{H}$ to be the angle between these subspaces).

Let $Vect_n(X)$ denote the set of isomorphism classes of n-dimensional vector bundles over X. Also, for any two spaces X,Y let [X,Y] denote the set of homotopy classes of maps $X \to Y$.

(2.5) Theorem. If X is a paracompact space, there is a natural one-one correspondence $Vect_n(X) \leftrightarrow [X,BGL_n]$.

We indicate briefly the way in which this correspondence is defined. Given $\pi : E \to X$ of dimension n, the paracompactness of X makes it possible to find a countable open cover $\{U_i\}_{i=1}^{\infty}$ of X and a subordinate partition of unity $\{\lambda_i\}_{i=1}^{\infty}$ and a continuous $\varphi_i : E|U_i \to R^n$ mapping each fiber isomorphically onto R^n. Express $\mathcal{H}$ as a countable orthogonal direct sum of copies of R^n and let $\psi_i : R^n \to \mathcal{H}$ be the inclusion of the ith summand. Then $\varphi : E \to \mathcal{H}$ defined by

$$\varphi(e) = \sum_{i=1}^{\infty} \lambda_i(\pi(e)) \cdot \psi_i \circ \varphi_i(e)$$

is continuous and sends each fiber $\pi^{-1}(x)$ linearly and one-one onto an n-dimensional subspace of $\mathcal{H}$. Thus $g_E(x) = \varphi(\pi^{-1}(x))$ defines a continuous map $g_E : X \to BGL_n$. This construction is then shown to set up the desired correspondence. For a careful treatment of (2.5), cf. [2, Chapter III].

As an example, let M be an n-dimensional manifold and choose a smooth imbedding $M \subset R^k$. Then, for each $x \in M$, $T_x(M)$ identifies canonically with an n-dimensional subspace of $R^k \subset \mathcal{H}$. This enables us to construct $\varphi : T(M) \to \mathcal{H}$ as desired. The corresponding map

$$g_{T(M)} : M \to BGL_n$$

is called the "Gauss map". The reader who needs to increase his comfort with these ideas is invited to prove that the above construction really does well define $g_{T(M)}$ up to homotopy.

(2.6) Theorem. Modulo torsion, the cohomology ring $H^*(BGL_n)$ is isomorphic to a polynomial ring $\mathbb{Z}[p_1,\ldots,p_{[n/2]}]$ where $p_i \in H^{4i}(BGL_n)$ is canonically defined. Thus, as algebras over R,

$$H^*(BGL_n; R) \cong R[p_1,\ldots,p_{[n/2]}] \quad .$$

For a proof, cf. [1,p.375]

(2.7) Definition. If $\pi : E \to X$ is an n-dimensional vector bundle over paracompact X, let $g_E : X \to BGL_n$ be the map, unique up to homotopy, given by (2.5). Then the ith (real) Pontryagin class of E is defined to be

$$p_i(E) = g_E^*(p_i) \in H^{4i}(X;R),$$

$i = 1,\ldots,[n/2]$. Furthermore, the graded subring

$$\mathrm{Pont}^*(E) = g_E^*(H^*(BGL_n;R)) \subset H^*(X;R)$$

is called the (real) Pontryagin ring of E.

We are now in a position to state the theorem whose proof is our first main aim.

(*) Theorem. If $E \subset T(M)$ is integrable and if the quotient bundle $Q = T(M)/E$ has fiber dimension q, then $\mathrm{Pont}^k(Q) = 0$ for $k > 2q$.

This is really a <u>global</u> integrability condition. Indeed, for $\pi : E \to X$ any n-dimensional bundle, set

$$p(E) = 1 + p_1(E) + \cdots + p_{[n/2]}(E) .$$

Because of the leading term 1 this is an invertible element of the ring $H^{**}(X;R)$ of formal infinite series $a_o + a_1 + \cdots + a_r + \cdots$, $a_i \in H^i(X;R)$. If $E' \subset E$ is a subbundle, the basic "duality" formula [3]

$$p(E/E') = p(E')^{-1} p(E)$$

holds and shows that the Pontryagin classes of E/E' depend only on the isomorphism classes of E and E' and <u>not</u> on the imbedding $E' \subset E$. Thus we can reformulate (*) as follows.

(*)′ Theorem. If $E \subset T(M)$ is a subbundle which is isomorphic to an integrable subbundle $E' \subset T(M)$, and if $Q = T(M)/E$, $q = \dim(Q)$, then $\mathrm{Pont}^k(Q) = 0$ for $k > 2q$.

Thus we have a topological obstruction to solving the global integrability problem (1.4). For some time this was the only known topological obstruction to global integrability. However, Shulman has observed (cf. §6) that the method of proof of (*) also shows that the Massey products of elements of $\mathrm{Pont}^*(Q)$ must vanish in suitably high dimensions under the hypothesis of integrability. Examples exist in which these new obstructions do not all vanish although $\mathrm{Pont}^*(Q)$ does vanish in dimensions $> 2q$.

References

1. A. Borel and F. Hirzebruch, Characteristic classes and homogeneous spaces, II, Am. J. Math., 81 (1959), pp. 315-382.

2. D. Husemoller, Fibre Bundles, McGraw-Hill, New York, 1966.

3. J. Milnor, Lectures on Characteristic Classes, (Notes by J. Stasheff), Mimeographed notes, Princeton Univ.

4. Munkres, Elementary Differential Topology, Annals of Math. Studies, No. 54, Princeton Univ. Press, Princeton, N. J., 1963.

3. De Rham cohomology. In order to prove (*) we will use an alternative definition of the Pontryagin classes in terms of differential forms. In order to formulate this definition we need a cohomology theory based on differential forms, the de Rham cohomology $H^*_{DR}(M)$. The present section is devoted to the definition of this theory and a sketch of its basic properties.

Let Λ^*_n be the associative algebra over $\mathbb{R}$ generated by elements $1, dx_1, \ldots, dx_n$ subject to the relations

1) 1 is a multiplicative identity

2) $dx_i \cdot dx_j = - dx_j \cdot dx_i$, $\forall\, i,j$.

Remark that a consequence of 2) is the relation

$$dx_i^2 = 0, \ \forall\, i .$$

Thus any monomial $dx_{i_1} \cdots dx_{i_r} = 0$ if $r > n$. This algebra has a natural grading by the degree of a monomial:

$$\Lambda^*_n = \bigoplus_{j=0}^{n} \Lambda^j_n \quad ,$$

where, of course, $\Lambda_n^o = R \cdot 1 \cong R$. Each Λ_n^j is a finite dimensional vector space over R.

(3.1) Definition. Let $U \subset R^n$ be an open subset. Let $A^k(U) = \{\omega : U \to \Lambda_n^k : \omega \text{ smooth}\}$. Then $A^*(U) = \bigoplus_{k=0}^{n} A^k(U)$ with its natural graded associative algebra structure is called the algebra of differential forms on U. Each $\omega \in A^k(U)$ is called a k-form on U.

Notice that each $f \in A^0(U)$ is simply a smooth real valued function on U. In general, $\omega \in A^k(U)$ can be written uniquely as

$$\omega = \sum_{1 \leq i_1 < \cdots < i_k \leq n} f_{i_1 \ldots i_k} dx_{i_1} \cdots dx_{i_k}$$

where each $f_{i_1 \ldots i_k}$ is a smooth real valued function on U.

(3.2) Definition. The exterior differential $d : A^0(U) \to A^1(U)$ is given by the formula $df = \sum_{j=1}^{n} \frac{\partial f}{\partial x_j} dx_j$. Generally, the exterior differential $d : A^k(U) \to A^{k+1}(U)$ is given by

$$d\Big(\sum_{i_1 < \cdots < i_k} f_{i_1 \ldots i_k} dx_{i_1} \cdots dx_{i_k}\Big)$$

$$= \sum_{i_1 < \cdots < i_k} d(f_{i_1 \ldots i_k}) dx_{i_1} \cdots dx_{i_k} .$$

Remark that if $x_i : U \to R$ is the ith coordinate function, then $d(x_i) = dx_i$.

(3.3) Exercises. 1) Prove that $d^2 = 0$. Indeed, show that this is simply a fancy way of saying that, for smooth real valued functions f on U, the mixed partials satisfy $\frac{\partial^2 f}{\partial x_i \partial x_j} = \frac{\partial^2 f}{\partial x_j \partial x_i}$, $\forall\ i,j$.

2) If $\omega \in A^p(U)$, $\eta \in A^q(U)$, prove that $d(\omega\eta) = d(\omega)\eta + (-1)^p \omega d(\eta)$.

(3.4) Definition. If $U \subset R^n$ and $V \subset R^m$ are open sets, $F : U \to V$ a smooth map, then define

$$F^* : A^0(V) \to A^0(U)$$

by $F^*(f) = f \circ F$. Define

$$F^*(dx_i) = d(F^*(x_i))$$

and extend this to a unique homomorphism of graded algebras

$$F^* : A^*(V) \to A^*(U).$$

Remark that $F^* \circ d = d \circ F^*$.

We would like to extend all of the above notions to any n-dimensional manifold M. Using local coordinate charts on M, one can define a smooth k-form on a coordinate neighborhood $U_\alpha \subset M$. ω_α is said to agree with ω_β if the smooth change of coordinates on $U_\alpha \cap U_\beta$ transforms ω_α into ω_β. If all of the k-forms ω_α agree, they are said to define a (smooth) k-form $\omega \in A^k(M)$. By the above it is clear that

$$d : A^k(M) \to A^{k+1}(M)$$

is still well-defined and that $d^2 = 0$.

(3.5) Definition. Let $Z^k_{DR}(M) = \{\omega \in A^k(M) : d\omega = 0\}$ and $B^k_{DR}(M) = \{\omega \in A^k(M) : \omega = d\eta$, some $\eta \in A^{k-1}(M)\}$. The de Rham cohomology of M is defined to be

$$H^*_{DR}(M) = \bigoplus_{k=0}^{n} H^k_{DR}(M)$$

where

$$H^k_{DR}(M) = \frac{Z^k_{DR}(M)}{B^k_{DR}(M)} .$$

Notice that a smooth map $F : M \to N$ induces $F^* : A^*(N) \to A^*(M)$ with $F^* \circ d = d \circ F^*$, hence induces

$$F^* : H^*_{DR}(N) \to H^*_{DR}(M) .$$

Clearly, $(id)^* = id$ and $(F \circ G)^* = G^* \circ F^*$, so de Rham cohomology defines a contravariant functor from the category of smooth manifolds and smooth maps to the category of graded R-algebras and graded R-algebra homomorphisms.

A basic property of de Rham cohomology is its homotopy invariance. This will be a consequence of the following lemma.

(3.6) **Lemma.** Let $\pi : M \times R \to M$ be the standard projection. Then $\pi^* : H^*_{DR}(M) \to H^*_{DR}(M \times R)$ is bijective.

Proof. Let $i_0 : M \to M \times R$ be defined by $i_0(x) = (x,0)$. Then $\pi \circ i_0 = id$, so $i_0^* \circ \pi^* = id : H^*_{DR}(M) \to H^*_{DR}(M)$. We have to show that $\pi^* \circ i_0^*$ is also an identity map.

Every $\omega \in A^k(M \times R)$ is a linear combination of k-forms of the following two types:

a) $\pi^*(\varphi) \cdot f(x,t)dt$, $\varphi \in A^{k-1}(M)$

b) $\pi^*(\varphi) \cdot f(x,t)$, $\varphi \in A^k(M)$.

Define $S : A^k(M \times R) \to A^{k-1}(M \times R)$ by

$$S(\pi^*(\varphi)\cdot f(x,t)) = 0$$

$$S(\pi^*(\varphi)\cdot f(x,t)dt) = \pi^*(\varphi)\cdot g$$

where

$$g(x,t) = \int_0^t f(x,u)du .$$

This is clearly well defined and a straightforward computation gives

$$(-1)^{k-1}(d\circ S - S\circ d) = id - \pi^*\circ i_0^*$$

on $A^k(M \times R)$. Thus $id - \pi^*\circ i_0^*$ maps $Ker(d)$ into $Im(d)$, hence induces the zero homomorphism in cohomology. This proves the assertion.

q.e.d.

An obvious induction proves the following important corollary.

(3.7) Corollary. (Poincaré Lemma).

$$H^k_{DR}(R^q) = H^k_{DR}(\text{point}) = \begin{cases} R, & k = 0 \\ 0, & k > 0 \end{cases} .$$

This corollary is usually formulated differently. Indeed, consider the sequence

$$0 \to R \xrightarrow{i} A^0(R^q) \xrightarrow{d} A^1(R^q) \xrightarrow{d} \cdots \xrightarrow{d} A^q(R^q) \to 0 .$$

Here, if $a \in R$ then $i(a)$ is the constant function a on R^q. Evidently

$$Im(i) = Ker\{d: A^0(R^q) \to A^1(R^q)\} .$$

Then, (3.7) is entirely equivalent to the assertion that this sequence is exact.

(3.8) Corollary. If $f_0, f_1 : M \to N$ are smooth and are smoothly homotopic, then $f_0^* = f_1^* : H^*_{DR}(N) \to H^*_{DR}(M)$.

Proof. Let $i_0, i_1 : M \to M \times R$ be defined by

$$i_0(x) = (x,0)$$

$$i_1(x) = (x,1).$$

If $F : M \times I \to N$ is the smooth homotopy between f_0 and f_1 it can be smoothly extended to $F : M \times R \to N$ and

$$f_0 = F \circ i_0$$

$$f_1 = F \circ i_1 .$$

Thus

$$f_0^* = i_0^* \circ F^*$$

$$f_1^* = i_1^* \circ F^*$$

and it will be enough to show $i_0^* = i_1^*$ in cohomology. Consulting the proof of (3.6) we see that $i_0^* = (\pi^*)^{-1}$ and that the same argument shows $i_1^* = (\pi^*)^{-1}$. Thus $i_0^* = i_1^*$. q.e.d.

Closely related to the above discussion is an important operation in de Rham cohomology called "integration along the fiber". We describe it briefly.

If M is oriented and n-dimensional, and if $A_0^n(M)$ denotes the space of n-forms with compact support, then each $\omega \in A_0^n(M)$ has a well defined integral $\int_M \omega \in R$ (cf.[3,p.118]). If the orientation of M is reversed, this integral receives the opposite sign. If M has boundary $\partial M \neq \phi$, all of our definitions make sense and one has the following basic theorem.

(3.9) Theorem. (Stokes) Let $i : \partial M \to M$ be the inclusion. Then $\int_M d\omega = \int_{\partial M} i^*\omega , \ \forall \omega \in A_0^{n-1}(M)$.

Let $\pi : E \to M$ be a smooth fiber bundle with fiber a compact manifold F. We assume that $\partial M \neq \phi$ and note that, if $\partial F \neq \phi$, then $\partial E \neq \phi$ and π restricts to a smooth fiber bundle $\pi^\partial : \partial E \to M$ with fiber ∂F. Let $e = \dim(E)$, $f = \dim(F)$.

(3.10) Theorem. For $\pi : E \to M$ as above, there is a canonical homomorphism

$$\pi_* : A^r(E) \to A^{r-f}(M), \quad r \geq 0 ,$$

which is zero for $r < f$ and satisfies

$$\pi_* \circ d + (-1)^{f+1} d \circ \pi_* = \pi_*^\partial \circ i^*$$

for $r \geq f$. Here, if $\partial E = \phi$, the right hand side is interpreted as 0.

Indeed, one can show the existence of a unique π_* satisfying the equation

$$\int_E \psi \cdot \pi^*(\varphi) = \int_M \pi_*(\psi) \cdot \varphi$$

for all $\psi \in A^r(E)$, $\varphi \in A_0^{e-r}(M)$. Then, for $\omega \in A^r(E)$, $\varphi \in A_0^{e-r-1}(M)$, we have $\int_M (\pi_* d\omega + (-1)^{f+1} d\pi_*\omega) \cdot \varphi = \int_M \pi_*(d\omega) \cdot \varphi$

$$+ (-1)^{f+1} \int_M \{d(\pi_*(\omega) \cdot \varphi) - (-1)^{r-f} \pi_*(\omega) \cdot d\varphi\}$$

$$= \int_M \pi_*(d\omega) \cdot \varphi + (-1)^r \int_M \pi_*(\omega) \cdot d\varphi$$

$$= \int_E d\omega \cdot \pi^*(\varphi) + (-1)^r \int_E \omega \cdot \pi^*(d\varphi)$$

$$= \int_E d(\omega \cdot \pi^*(\varphi)) = \int_{\partial E} i^*(\omega) \cdot \pi^{\partial *}(\varphi)$$

$= \int_M \pi_*^\partial i^*(\omega) \cdot \varphi$. This proves the desired identity.

If $\partial E = \phi$, (3.10 says that π_* either commutes or anticommutes with d, hence, that it induces a homomorphism

$$\pi_* : H_{DR}^r(E) \to H_{DR}^{r-f}(M) .$$

This homomorphism can also be produced in the standard cohomology of fiber bundles by a spectral sequence technique (cf. [1, §8]).

Returning to the study of the de Rham cohomology functor, we remark first that $H^*_{DR}(M)$ has the structure of a graded algebra over $\mathbb{R}$. Indeed, the relation

$$d(\omega \cdot \eta) = d(\omega) \cdot \eta + (-1)^{\deg(\omega)} \omega \cdot d\eta ,$$

assigned in (3.3) as an exercise, implies that $Z^*_{DR}(M)$ is a subring of $A^*(M)$ and that $B^*_{DR}(M)$ is an ideal in $Z^*_{DR}(M)$. Thus $H^*_{DR}(M)$ has a graded algebra structure inherited from that of $A^*(M)$.

Let $\check{H}^*(M;\mathbb{R})$ denote the standard Čech cohomology of M with coefficients $\mathbb{R}$. This is a graded algebra under the cup product. The following theorem is of fundamental importance for relating topological invariants of manifolds to their differential geometric properties.

(3.11) Theorem (de Rham) There is a canonical isomorphism of graded algebras

$$\theta_M : H^*_{DR}(M) \longrightarrow \check{H}^*(M;\mathbb{R}).$$

Furthermore, if $f : M \to N$ is smooth, then the diagram

$$\begin{array}{ccc} H^*_{DR}(M) & \xrightarrow{\theta_M} & \check{H}^*(M;\mathbb{R}) \\ \uparrow f^* & & \uparrow f^* \\ H^*_{DR}(N) & \xrightarrow{\theta_N} & \check{H}^*(N;\mathbb{R}) \end{array}$$

is commutative.

We sketch a proof of (3.11) due to André Weil [4]. Let U be an open cover of M and for each pair (p,q) of nonnegative integers, set

$$K^{p,q}(U) = \check{C}^p(U;A^q) ,$$

the pth Čech cochain module with values in the q-forms. Precisely, each $c \in K^{p,q}(U)$ is a function which to each ordered $(p+1)$-tuple $(U_{\alpha_0},\ldots,U_{\alpha_p})$ of elements of U assigns a q-form

$$c_{\alpha_0\alpha_1\ldots\alpha_p} \in A^q(U_{\alpha_0} \cap \ldots \cap U_{\alpha_p})$$

(interpreted as 0 if $U_{\alpha_0} \cap \ldots \cap U_{\alpha_p} = \emptyset$). The Čech coboundary

$$\delta : K^{p,q}(U) \to K^{p+1,q}(U)$$

is given by the formula

$$\{\delta(c)\}_{\alpha_0\ldots\alpha_{p+1}} = \sum_{i=0}^{p+1} (-1)^i \varphi_{\alpha_i}(c_{\alpha_0\ldots\hat{\alpha}_i\ldots\alpha_{p+1}})$$

where

$$\varphi_{\alpha_i} : A^q(U_{\alpha_0} \cap \ldots \cap \hat{U}_{\alpha_i} \cap \ldots \cap U_{\alpha_{p+1}}) \to A^p(U_{\alpha_0} \cap \ldots \cap U_{\alpha_{p+1}})$$

is the restriction map. As usual, $\delta^2 = 0$. The de Rham coboundary

$$d : K^{p,q}(U) \to K^{p,q+1}(U)$$

is given by

$$\{d(c)\}_{\alpha_0\ldots\alpha_p} = d(c_{\alpha_0\ldots\alpha_p}) .$$

Clearly $d\delta = \delta d$ on $K^{**}(U)$. Set

$$D' = \delta$$
$$D'' = (-1)^p d \quad (\text{on } K^{p,q}(U))$$
$$D = D' + D''$$

and remark that $D^2 = 0$. Set

$$K^n(U) = \bigoplus_{p+q=n} K^{p,q}(U)$$

and remark that

$$D : K^n(U) \to K^{n+1}(U).$$

One can also define a multiplication

$$K^{p,q}(U) \times K^{r,s}(U) \to K^{p+r,q+s}(U)$$

by

$$(\omega\eta)_{\alpha_0\ldots\alpha_{p+r}} = (-1)^{qr}\, \omega_{\alpha_0\ldots\alpha_p} \cdot \eta_{\alpha_p\ldots\alpha_{p+r}}$$

on their common domain. This also defines a multiplication

$$K^n(U) \times K^m(U) \to K^{n+m}(U) .$$

Note that if $\omega \in K^{p,q}$, $\eta \in K^{r,s}$, then

$$D'(\omega\eta) = \delta((-1^{rq}\omega \cdot \eta)$$

where $\cdot$ indicates exterior multiplication of forms, hence

$$D'(\omega\eta) = (-1)^{rq} \{\delta(\omega).\eta + (-1)^p\omega \cdot \delta(\eta)\}$$

$$= (-1)^{rq}(-1)^{rq}D'(\omega)\eta + (-1)^{p+rq}(-1)^{q(r+1)}\omega\, D'(\eta)$$

$$= D'(\omega)\eta + (-1)^{p+q}\omega\, D'(\eta) .$$

Likewise,

$$D''(\omega\eta) = (-1)^{p+r}d((-1)^{rq}\omega \cdot \eta)$$

$$= (-1)^{p+r+rq}\{d(\omega)\cdot\eta + (-1)^q\omega\cdot d\eta\}$$

$$= (-1)^{p+r+rq+p+(q+1)r}D''(\omega)\eta + (-1)^{p+q+r+rq+r+rq}\omega\, D''(\eta)$$

$$= D''(\omega)\eta + (-1)^{p+q}\omega\, D''(\eta) .$$

Thus $D : K^*(U) \to K^*(U)$ is an antiderivation.

Note that the kernel of

$$D' : K^{0,q}(U) \to K^{1,q}(U)$$

is exactly $A^q(M)$ and that the kernel of

$$D'' : K^{p,0}(U) \to K^{p,1}(U)$$

is eactly $\check{C}^p(U;R)$. We define the homomorphisms

$$\alpha : A^*(M) \to K^*(U)$$

$$\beta : \check{C}^*(U;R) \to K^*(U)$$

to be the inclusion maps. These are homomorphisms of cochain complexes and homomorphisms of graded algebras.

(3.12) **Definition.** U is called a simple covering if every finite nonempty intersection of elements of U is a contractible set.

(3.13) **Lemma.** Every open cover of M admits a refinement which is a simple covering.

Proof. Put a Riemannian metric on M and choose the refinement such that each element is a geodesically convex set (cf. [2,p.39]). Any finite nonempty intersection is then again geodesically convex, hence contractible.

q.e.d.

Since we are going to pass to the limit over open coverings, (3.13) allows us to assume that U is a simple covering.

Consider the first quadrant array

$$
\begin{array}{ccccccccccc}
\vdots & & \vdots & & \vdots & & & & \vdots & & \\
\uparrow d & & \uparrow D'' & & & & & & \uparrow D'' & & \\
A^q(M) & \xrightarrow{\alpha} & K^{0q}(U) & \xrightarrow{D'} & \cdots & & & \xrightarrow{D'} & K^{pq}(U) & \xrightarrow{D'} & \cdots \\
\uparrow d & & \uparrow D'' & & \vdots & & & & \uparrow D'' & & \\
\vdots & & \vdots & & \vdots & & & & \vdots & & \\
\uparrow d & & \uparrow D'' & & & & & & & & \\
A^1(M) & \xrightarrow{\alpha} & K^{01}(U) & \xrightarrow{D'} & \cdot & & & & \cdot & & \\
\uparrow d & & \uparrow D'' & & \uparrow D'' & & & & \uparrow D'' & & \\
A^0(M) & \xrightarrow{\alpha} & K^{00}(U) & \xrightarrow{D'} & K^{10}(U) & \xrightarrow{D'} & \cdots & \xrightarrow{D'} & K^{p0}(U) & \xrightarrow{D'} & \cdots \\
 & & \uparrow \beta & & \uparrow \beta & & & & \uparrow \beta & & \\
 & & \check{C}^0(U;R) & \xrightarrow[\delta]{} & \check{C}^1(U;R) & \xrightarrow[\delta]{} & \cdots & \xrightarrow[\delta]{} & \check{C}^p(U;R) & \xrightarrow[\delta]{} & \cdots
\end{array}
$$

(3.14) Lemma. In the above diagram, the rows are exact. If U is a simple covering the columns are also exact.

Proof. Let $\lambda = \{\lambda_\alpha\}$ be a smooth partition of unity subordinate to the open cover $U = \{U_\alpha\}$. If $c \in \check{C}^p(U;A^q)$, define $L(c) \in \check{C}^{p-1}(U;A^q)$ by

$$\{L(c)\}_{\alpha_0 \ldots \alpha_p} = \sum_\alpha \lambda_\alpha \cdot c_{\alpha\alpha_0 \ldots \alpha_{p-1}}$$

where each $\lambda_\alpha \cdot c_{\alpha\alpha_0 \ldots \alpha_{p-1}}$ is interpreted in the obvious way as an element of $A^q(U_{\alpha_0} \cap \ldots \cap U_{\alpha_{p-1}})$. If $D'(c) = 0$, then a straightforward computation gives $c = D'L(c)$, which proves exactness of the rows. If U is simple, then the columns are exact by (3.7). q.e.d.

Assume in all that follows that U is simple.

(3.15) Lemma. In cohomology, $\alpha^* : H^*_{DR}(M) \to H^*(K^*, D)$ and

$\beta^* : H^*(U;R) \to H^*(K^*,D)$ are one-one.

The proof of this is a trivial exercise.

(3.16) Lemma. In cohomology; α^* and β^* are onto.

Proof. Let $z \in K^r(U)$ such that $D(z) = 0$. Write

$$z = \sum_{p=0}^{n} z_p , \qquad z_p \in K^{p,n-p}(U) .$$

Since $D(z) = 0$, we have in particular that $D''(z_0) = 0$. By exactness of columns there is $u \in K^{0,n-1}$ such that $D''(u) = z_0$. Then $z - D(u)$ has component zero in $K^{0,n}(U)$. Repeating this process we finally produce a D-cocycle w lying in $K^{n,0}(U)$ cohomologous to z. But $D(w) = 0$ implies that

$$D'(w) = 0 \quad \text{and} \quad D''(w) = 0 \quad ,$$

hence $w = \beta(v)$ where $v \in \check{C}^n(U;R)$ is a δ-cocycle. This shows β^* onto. An entirely parallel argument using exactness of the rows proves α^* onto.

q.e.d.

(3.11) follows easily. Indeed

$$\alpha^* : H^*_{DR}(M) \to H^*(K^*(U),D)$$

is an isomorphism for every simple covering U, hence the right hand side has already attained its limit value. Since

$$\beta^* : \check{H}^*(U;R) \to H^*(K^*(U),D)$$

is also an isomorphism, the left hand side of this relation has likewise reached its limit value $\check{H}^*(M;R)$. Thus, set $\theta_M = (\beta^*)^{-1} \circ \alpha^*$, the desired isomorphism of graded algebras. We leave it to the reader to check the naturality of θ_M relative to smooth maps.

References

1. A. Borel and F. Hirzebruch, Characteristic classes and homogeneous spaces, I, Am. J. Math., 80 (1958), pp. 458-538.

2. S. Helgason, Differential Geometry and Symmetric Spaces, Academic Press, New York, 1962.

3. S. Sternberg, Lectures on Differential Geometry, Prentice-Hall, Englewood Cliffs, N. J., 1964.

4. A. Weil, Sur les théorèmes de de Rham, Comment. Math. Helv., 26 (1952), pp. 119-145.

4. Connections on vector bundles. Let $\pi : E \rightarrow M$ be a smooth vector bundle and let $\Gamma(E)$ denote the set of smooth cross sections. That is, the elements $s \in \Gamma(E)$ are smooth functions

$$s : M \rightarrow E$$

such that $\pi \circ s = \mathrm{id}$. Clearly $\Gamma(E)$ is a vector space over $\mathbb{R}$ and a module over the algebra $\mathfrak{F}$ of all smooth real valued functions on M. Let $\mathfrak{X}$ denote $\Gamma(T(M))$.

(4.1) Definition. A connection on E is an $\mathbb{R}$-bilinear map

$$\nabla : \mathfrak{X} \times \Gamma(E) \rightarrow \Gamma(E)$$

written $\nabla(X,s) = \nabla_X(s)$, such that

1) $\nabla_X(fs) = X(f)s + f\,\nabla_X(s)$

2) $\nabla_{fX}(s) = f\,\nabla_X(s)$

for all $X \in \mathfrak{X}$, $s \in \Gamma(E)$, $f \in \mathfrak{F}$.

While the motivation of this definition may not at first be evident, a little reflection shows that ∇_X is a kind of directional derivative which differentiates smooth sections of E in X-directions. Indeed, property 2) in the definition shows that at any point of M, $\nabla_X(s)$ depends (for fixed s) only on the value of X at that point. Thus if $X_0 \in T_{x_0}(M)$, ∇_{X_0} is a well defined operator transforming smooth sections of E defined near x_0 into vectors in the fiber $\pi^{-1}(x_0)$. Property 1) says that this is a first order differential operator.

(4.2) Lemma. There exists a connection ∇ on E.

Proof. Let $\{U_\alpha\}$ be an open cover of M such that for each α there is a smooth frame $s_\alpha = (s_\alpha^1,\ldots,s_\alpha^q)$ defined on U_α. Then each $s_\alpha^i \in \Gamma(E|U_\alpha)$ and for each $x \in U_\alpha$, $\{s_\alpha^1(x),\ldots,s_\alpha^q(x)\}$ is a basis for $\pi^{-1}(x)$. Let $\{\lambda_\alpha\}$ be a smooth partition of unity subordinate to this cover.

Over U_α define a connection ∇^α on $E|U_\alpha$ by

$$\nabla_X^\alpha(s_\alpha^i) = 0, \quad \forall\, i, \forall\, X ,$$

which extends to arbitrary sections s by the property 1) in (4.1).

$\nabla = \sum_\alpha \lambda_\alpha \nabla^\alpha$ then makes sense and satisfies (4.1). q.e.d.

We may consider ∇ as an $\mathfrak{F}$-linear map

$$\nabla : \mathfrak{X} \to \mathrm{Hom}_R(\Gamma(E),\Gamma(E)) .$$

Both of these $\mathfrak{F}$-modules are also Lie algebras over R. Generally, ∇ is <u>not</u> a homomorphism of Lie algebras and the measure of its failure to be such is an important geometric quantity.

(4.3) Definition. The curvature k of the connection ∇ is the function

$$k : \mathfrak{X} \times \mathfrak{X} \to \mathrm{Hom}_R(\Gamma(E), \Gamma(E))$$

given by

$$k(X,Y) = [\nabla_X, \nabla_Y] - \nabla_{[X,Y]}$$

$$= \nabla_X\nabla_Y - \nabla_Y\nabla_X - \nabla_{[X,Y]}.$$

Remark that $k(X,Y) = - k(Y,X)$.

(4.4) Lemma. If $f,g,h \in \mathfrak{F}$, $X,Y \in \mathfrak{X}$, and $s \in \Gamma(E)$, then $k(fX,gY)(hs) = fgh \cdot k(X,Y)(s)$.

The proof is a direct verification which we leave to the reader. As a consequence, consider a smooth frame $(s_\alpha^1,\dots,s_\alpha^q)$ for $E|U_\alpha$ and note that there is a $q \times q$ matrix k^α of 2-forms, $k^\alpha = (k_{ij}^\alpha)$ such that

$$k(X,Y)(s_\alpha^i) = \sum_{j=1}^{q} k_{ji}^\alpha(X,Y)s_\alpha^j \quad .$$

Here, of course, we use the identification of $A^q(M)$ with the set of alternating $\mathfrak{F}$-multilinear q-forms on $\mathfrak{X}$ which is standard in differential geometry. The fact that $k_{ji}^\alpha(X,Y) = -k_{ji}^\alpha(Y,X)$ together with $\mathfrak{F}$-bilinearity then identifies k_{ji}^α as an element of $A^2(U_\alpha)$.

Referring to the definition (2.1) of vector bundles, we note that, on $U_\alpha \cap U_\beta$, k^α and k^β are related by

$$k^\alpha = g_{\alpha\beta}k^\beta g_{\alpha\beta}^{-1} \quad .$$

The proof of this crucial formula is an elementary exercise using (4.4).

We also note that on U_α there is a $q \times q$ matrix θ^α of 1-forms, $\theta^\alpha = (\theta_{ij}^\alpha)$, such that

$$\nabla_X(s_\alpha^i) = \sum_{j=1}^{q} \theta_{ij}^\alpha(X)s_\alpha^i \quad .$$

The following basic structural equation will be found proven in various books on differential geometry (e.g., cf.[1, p.118]).

(4.5) Lemma. $d\theta^{\alpha} - \theta^{\alpha} \cdot \theta^{\alpha} = k^{\alpha}$.

Here, of course, the matrix product $\theta^{\alpha} \cdot \theta^{\alpha}$ involves the usual multiplication of 1-forms in the entries.

As a final remark about connections, we define the notion of parallel translation. If $\sigma : [0,1] \to M$ is a smooth curve, one defines the notion of a "section of E along σ" as follows.

(4.6) Definition. A section of E along σ is a smooth curve $s : [0,1] \to E$ such that $\pi \circ s = \sigma$.

If s is a section of E along σ, the connection ∇ defines the "covariant derivative of s along σ", a new section $\frac{Ds}{dt}$ of E along σ. If σ is regular (hence locally one-one) it is an easy exercise to extend s locally to a section of E and show that $\nabla_{\dot{\sigma}(t)}(s)$ does not depend on this extension, hence defines $\frac{Ds}{dt}$. To handle general σ, proceed as in [2, §8] .

(4.7) Definition. A section s of E along σ is parallel iff $\frac{Ds}{dt} \equiv 0$.

Given $e_0 \in T_{\sigma(0)}(M)$, the parallel translate of e_0 along σ will be the unique section e of E along σ satisfying

1) $e(0) = e_0$

2) e is parallel.

The existence and uniqueness of e is an elementary consequence of the theory of ordinary differential equations.

References

1. S. Kobayashi and K. Nomizu, Foundations of Differential Geometry, Vol. I, Interscience Publishers, New York. N. Y., 1963.

2. J. Milnor, Morse Theory, Princeton University Press, Princeton, N. J., 1963.

5. The Pontryagin classes. We suppose given a smooth vector bundle

$$\pi : E \to M$$

and a connection ∇ on E. Let k be the curvature of ∇. Let $g\ell_q$ denote the Lie algebra of $GL_q = GL(q,R)$. Thus $g\ell_q$ is the set of all q x q real matrices, while GL_q is the set of nonsingular q x q real matrices.

(5.1) Definition. A polynomial function

$$\varphi : g\ell_q \to R$$

is called invariant if

$$\varphi(m) = \varphi(gmg^{-1}), \forall g \in GL_q, \forall m \in g\ell_q .$$

Since sums and products of invariant polynomials are invariant, these polynomials form an algebra over R.

There are certain basic invariant polynomials $\Sigma_0, \Sigma_1, \Sigma_2, \ldots$ which are defined as follows.

$$\Sigma_p(m) = \text{trace } (m^p).$$

The following purely algebraic theorem will be of use.

(5.2) Theorem. The algebra of inv riant polynomials has $\{\Sigma_i\}_{i=0}^{\infty}$ as a set of generators.

For a proof of (5.2), cf. the Appendix.

Now recall the curvature matrices k^{α} of 2-forms defined on U_{α}. If φ is an invariant polynomial of degree r, then $\varphi(k^{\alpha}) \in A^{2r}(U_{\alpha})$. Furthermore, on $U_{\alpha} \cap U_{\beta}$,

$$\varphi(k^{\alpha}) = \varphi(g_{\alpha\beta} k^{\beta} g_{\alpha\beta}^{-1}) = \varphi(k^{\beta}).$$

Thus these 2r-forms fit together to give a well defined element $\varphi(k) \in A^{2r}(M)$. Note that this form depends only on the curvature k, not on choices of local trivializations of E.

(5.3) Lemma. $d(\varphi(k)) = 0$.

Proof. Let $x_0 \in M$ be any point, U_{α} an open neighborhood of x_0 which trivializes E and on which there is a smooth spherical coordinate system centered at x_0. Let s_0 be any frame for $\pi^{-1}(x_0)$ and spread this out to a smooth frame s_{α} over U_{α} by parallel translations along the radial curves of the spherical coordinate system. Then, for all $X \in T_{x_0}(M)$, $\theta^{\alpha}(X) = 0$. (i.e., every θ^{α}_{ij} is a 1-form which vanishes at x_0). On U_{α},

$$d(\varphi(k)) = d(\varphi(k^{\alpha})) = d(\varphi(d\theta^{\alpha} - \theta^{\alpha} \cdot \theta^{\alpha})) .$$

But $\varphi(d\theta^{\alpha} - \theta^{\alpha} \cdot \theta^{\alpha})$ is a sum of monomials of forms in which either two of the forms vanish at x_0 or all of the forms are $d\theta^{\alpha}_{ij}$'s. Thus

$$\{d(\varphi(k))\}_{x_0} = 0 .$$

But x_0 was an arbitrary point of M, hence $d(\varphi(k)) = 0$. q.e.d.

(5.4) Proposition. $[\varphi(k)] \in H^{2r}_{DR}(M)$ is independent of the choice of connection.

Proof. Let ∇^0 and ∇^1 be two connections on E. Let k^0 and k^1 be the corresponding curvatures. In a natural way,

$$\pi \times id : E \times R \longrightarrow M \times R$$

defines a smooth vector bundle over $M \times R$ with the same fiber dimension as E. Define a connection $\tilde{\nabla}$ on this bundle as follows. On sections s which are constant in R - directions, let $\tilde{\nabla}_{\partial/\partial t}(s) = 0$. If $X \in T_{(x,t)}(M \times \{t\})$, define

$$\tilde{\nabla}_X(s) = (1-t)\nabla^0_X(s) + t\,\nabla^1_X(s) .$$

Every section is a function-linear combination of sections constant in R-directions, hence $\tilde{\nabla}$ extends to all sections by use of property 1) in (4.1). It is easy to check that $\tilde{\nabla}$ is a connection. Let $\tilde{k}$ be its curvature and consider $[\varphi(\tilde{k})] \in H^{2r}_{DR}(M \times R)$. Let i_0 and $i_1 : M \to M \times R$ be as usual. Then $i_0^* = i_1^*$ in cohomology by the proof of (3.6), and

$$i_0^*[\varphi(\tilde{k})] = [\varphi(k^0)]$$

$$i_1^*[\varphi(\tilde{k})] = [\varphi(k^1)] .$$

q.e.d.

(5.5) Definition. If φ is an invariant polynomial of degree r, $\varphi(E) = [\varphi(k)] \in H^{2r}_{DR}(M)$ is called the Pontryagin class corresponding to φ. The set of all such Pontryagin classes is the graded subalgebra

$$\mathrm{Pont}^*(E) \subset H^*_{DR}(M)$$

called the Pontryagin algebra of E.

(5.6) Proposition. $\mathrm{Pont}^j(E) = 0$ if j is not divisible by 4.

Proof. From the definition it is clear that this is 0 if j is not divisible by 2. To obtain the finer result, we put a smoothly varying positive definite inner product $< , >$ on the fibers of E. This is easily done locally and then pieced together with a smooth partition of unity. Next we want a connection ∇ on E such that

$$\nabla_X < s_1,s_2 > = < \nabla_X s_1,s_2 > + < s_1,\nabla_X s_2 > ,$$

$\forall X \in \mathfrak{X}$, $\forall s_1, s_2 \in \Gamma(E)$. On U_α choose a smooth orthonormal frame s_α. This is easy to do by applying Gram-Schmidt orthonormalization to any smooth frame. Then define ∇^α on U_α by requiring

$$\nabla^\alpha_X(s^i_\alpha) = 0, \quad i = 1,\ldots,q.$$

It is easy to check that ∇^α has the required property. Piece these local connections together by a smooth partition of unity and obtain ∇ as desired.

The curvature k of this connection has the important property

$$< k(X,Y)s_1,s_2 > = < s_1,-k(X,Y)s_2 > .$$

The proof of this is left as an elementary exercise. One immediate consequence is that, if on U_α we choose s_α to be a smooth orthonormal frame, then the matrix k^α is antisymmetric. Thus also, if $r > 0$ is odd, then $(k^\alpha)^r$ is antisymmetric, hence

$$\Sigma_r(k^\alpha) = \text{trace } ((k^\alpha)^r) = 0.$$

Thus $\Sigma_r(k) = 0$. But by (5.2) every invariant φ of odd degree is a linear combination of polynomials of the form $\Sigma_{i_1}\Sigma_{i_2}\cdots\Sigma_{i_m}$ in which some of the indices i_j are odd. Thus, $\deg(\varphi)$ odd implies $\varphi(k) = 0$. q.e.d.

This proposition enables us to define the total Pontryagin class of E to be

$$p(E) = 1 + p_1(E) + \cdots + p_{[q/2]}(E)$$

$$= [\det(I + \frac{\sqrt{-1}}{2\pi} k)]$$

where each $p_j(E) \in H^{4j}_{DR}(M)$. No imaginary components arise because of (5.6). These classes $1, p_1(E),\ldots,p_{[q/2]}(E)$ actually generate the whole algebra

$\mathrm{Pont}^*(E)$.

Note that this definition gives an easy proof of the formula $p(E \oplus E') = p(E) \cdot p(E')$.

For the case of complex vector bundles, Chern classes can be defined in complex de Rham cohomology $H^*_{DR}(M;C)$ by exactly the same techniques (using complex smooth p-forms $A^p_C(M)$). If E has complex dimension n, one obtains $c(E) = 1 + c_1(E) + \cdots + c_n(E)$, $c_i(E) \in H^{2i}_{DR}(M;C)$, and again $c(E \oplus E') = c(E) \cdot c(E')$.

For complex analytic bundles on a complex analytic manifold, holomorphic connections do not generally exist, although they can be defined locally. As a consequence, if one insists on using these local holomorphic connections there is generally no way of producing forms in the de Rham complex representing Chern classes, but it is possible to define in this way representative cocycles in the double complex $\check{C}^*(U; A^*_C(M)) = K^*(U)$. We sketch this construction.

Let $U = \{U_\alpha\}$ be a simple covering of M such that each U_α is a trivializing neighborhood for the complex analytic vector bundle E. One defines a holomorphic connection ∇^α on $E|U_\alpha$ by defining it as trivial on a holomorphic frame field. One then obtains a curvature form k^α on U_α and, if φ is an invariant polynomial of degree r on $g\ell(q,C)$, $\varphi(k^\alpha) \in A^{2r}_C(U_\alpha)$. Thus we define $\varphi^0 \in \check{C}^0(U;A^{2r}_C)$ by

$$\varphi^0_\alpha = \varphi(k^\alpha).$$

On $U_\alpha \cap U_\beta$ we have the two connections ∇^α and ∇^β . Thus on $(U_\alpha \cap U_\beta) \times C$ put the connection $(1-z)\nabla^\alpha + z\nabla^\beta$ defined analogously to the $\tilde{\nabla}$ in the proof of (5.4). Let $k^{\alpha\beta}$ be the corresponding curvature form, $\varphi(k^{\alpha\beta}) \in A^{2r}((U_\alpha \cap U_\beta) \times C)$. Let Δ^p denote the standard p-simplex.

Then

$$(U_\alpha \cap U_\beta) \times \Delta^1 \subset (U_\alpha \cap U_\beta) \times K \subset (U_\alpha \cap U_\beta) \times C$$

and the projection

$$\pi^{\Delta^1}: (U_\alpha \cap U_\beta) \times \Delta^1 \to U_\alpha \cap U_\beta$$

defines (via integration along the fiber)

$$\pi_*^{\Delta^1}: \check{C}^1(U;A_C^{2r}) \to \check{C}^1(U;A_C^{2r-1}).$$

Define $\qquad \varphi^1 \in \check{C}^1(U;A_C^{2r-1})$ by

$$\varphi^1_{\alpha\beta} = \pi_*^{\Delta^1}(\varphi(k^{\alpha\beta})).$$

On $U_\alpha \cap U_\beta \cap U_\gamma$ we work with three connections and the convex combination of these over $(U_\alpha \cap U_\beta \cap U_\gamma) \times C^2$ and we use

$$\pi^{\Delta^2}: (U_\alpha \cap U_\beta \cap U_\gamma) \times \Delta^2 \to U_\alpha \cap U_\beta \cap U_\gamma$$

to define $\qquad \varphi^2 \in \check{C}^2(U;A_C^{2r-2})$ by

$$\varphi^2_{\alpha\beta\gamma} = \pi_*^{\Delta^2}(\varphi(k^{\alpha\beta\gamma})) .$$

Continuing in this way, we produce

$$\bar{\varphi} = (\epsilon_0\varphi^0, \epsilon_1\varphi^1, \ldots, \epsilon_n\varphi^n) \in K^{2r}(U).$$

Here each $\epsilon_i = (-1)^{[(i+1)/2]}$. This makes $\bar{\varphi}$ a cocycle because of the relation (3.10)

$$\pi_*^{\Delta^p} \circ d + (-1)^{p+1} d \circ \pi_*^{\Delta^p} = \pi_*^{\partial\Delta^p} \circ i^* = \sum_{j=0}^{p} (-1)^j \pi_*^{\Delta^{p-1}(j)}$$

where $\Delta^{p-1}(j)$ is the jth face of Δ^p . The last equality is by the combinatorial version of Stoke's Theorem [1, p. 109]. Since $d\varphi(k^{\alpha_0 \ldots \alpha_p}) = 0$, the above relation guarantees that

$$d(\varphi^0) = 0$$

$$d(\varphi^p) = (-1)^{p+1}\delta(\varphi^{p-1}), \quad p > 0 \ .$$

Thus $\bar{\varphi}$ is a cocycle and we obtain the Chern class

$$\varphi(E) = [\bar{\varphi}] \in H^{2r}(K^*(U)) = H^{2r}_{DR}(M;C).$$

We might further note that, by the definition of integration along the fiber, the cochains φ^p will be 0 for $p > r$ (i.e., below the diagonal in $K^{**}(U)$).

References

1. S. Sternberg, Lectures on Differential Geometry, Prentice-Hall, Englewood Cliffs, N. J., 1964.

6. Characteristic classes and integrability

We here suppose that E is an integrable subbundle of $T(M)$. Let $Q = T(M)/E$ with fiber dimension q. Thus we assume the hypotheses of (*) in § 2 and must prove

$$\mathrm{Pont}^k(Q) = 0, \quad k > 2q \ .$$

Since E is integrable, (1.3)′ implies that all $X, Y \in \Gamma(E)$ must have $[X,Y] \in \Gamma(E)$. If $Z \in \Gamma(Q)$, then $Z = \pi(\tilde{Z})$, some $\tilde{Z} \in \mathfrak{X}$, where $\pi : T(M) \to Q$ is the canonical projection. $\tilde{Z}$ is well defined modulo elements of $\Gamma(E)$. Thus for $X \in \Gamma(E)$ and $Z \in \Gamma(Q)$,

$$\nabla_X(Z) = \pi[X, \tilde{Z}]$$

is well defined. This is clearly R-bilinear as a map

$$\nabla : \Gamma(E) \times \Gamma(Q) \longrightarrow \Gamma(Q)$$

and satisfies

1) $\nabla_X(fZ) = X(f)Z + f\nabla_X(Z)$,

2) $\nabla_{fX}(Z) = f\,\nabla_X(Z)$,

as is easily verified. This satisfies the definition of a connection on Q except that the variable X is restricted to range over $\Gamma(E) \subset \mathfrak{X}$. In order to complete ∇ to a connection, we use a Riemannian metric on T(M) to split this bundle into the direct sum of E and the orthogonal complement bundle to E in T(M). This complement is isomorphic to Q, so by the choice of Riemannian metric we have obtained an isomorphism

$$T(M) \cong E \oplus Q .$$

Let $\overline{\nabla}$ be any connection on Q. For $X \in \mathfrak{X} = \Gamma(E) \oplus \Gamma(Q)$, write $X = X_E + X_Q$. Then define

$$\nabla_X(Z) = \nabla_{X_E}(Z) + \overline{\nabla}_{X_Q}(Z),$$

$\forall\, Z \in \Gamma(Q)$. Thus if $X \in \Gamma(E) \subset \mathfrak{X}$, $X_Q = 0$ and we obtain the previous formula for $\nabla_X(Z)$. It is trivial to check that this defines a connection on Q.

(6.1) **Definition.** A basic connection ∇ on Q is one such that

$$\nabla_X(Z) = \pi[X,\tilde{Z}],$$

$\forall\, X \in \Gamma(E)$, where $\tilde{Z} \in \mathfrak{X}$ is such that $\pi(\tilde{Z}) = Z$.

We have proven

(6.2) **Lemma.** Under the assumption that E is integrable, there exists a basic connection on Q .

(6.3) Lemma. Let ∇ be a basic connection on Q, k the curvature of ∇. Then $k(X,X') \equiv 0$, $\forall\, X,X' \in \Gamma(E)$.

Proof. Let $Z \in \Gamma(Q)$ and $\widetilde{Z} \in \mathfrak{X}$ with $\pi(\widetilde{Z}) = Z$. Then

$$k(X,X')(Z) = \nabla_X\nabla_{X'}(Z) - \nabla_{X'}\nabla_X(Z) - \nabla_{[X,X']}(Z)$$

$$= \nabla_X(\pi[X',\widetilde{Z}]) - \nabla_{X'}(\pi[X,\widetilde{Z}]) - \pi[[X,X'],\widetilde{Z}] .$$

But we can choose

$$\widetilde{\pi[X,\widetilde{Z}]} = [X,\widetilde{Z}]$$

$$\widetilde{\pi[X',\widetilde{Z}]} = [X',\widetilde{Z}]$$

so

$$k(X,X')(Z) = \pi[X,[X',\widetilde{Z}]] - \pi[X',[X,\widetilde{Z}]] - \pi[[X,X'],\widetilde{Z}] = \pi(0) = 0$$

by the Jacobi identity. q.e.d.

(6.4) Lemma. Let $U_\alpha \subset M$ be a simultaneously trivializing neighborhood for Q and E, s_α a smooth frame for Q over U_α. Let $I_\alpha(E)$ be the ideal in $A^*(U_\alpha)$ generated by those 1-forms which vanish on $\Gamma(E|U_\alpha)$. Let k^α be the curvature matrix associated to the frame s_α by a basic connection. Then each $k^\alpha_{ij} \in I_\alpha(E)$.

Proof. Over U_α, E can be described as the set of tangent vectors on which certain 1 forms $\theta_1,\dots,\theta_q$ vanish, these 1-forms being linearly independent at each point of U_α. In particular, $I_\alpha(E)$ is generated by $\theta_1,\dots,\theta_q$. Complete these to a basis of 1-forms by $\theta_{q+1},\dots,\theta_n$. These last restrict to a basis of E^*_p, $\forall\, p \in U_\alpha$. Consider a nontrivial form

$$\omega = \sum_{q+1 < i < j} g_{ij}\theta_i \cdot \theta_j .$$

Clearly there are $X, X' \in \Gamma(E|U_\alpha)$ such that

$$\omega(X, X') \not\equiv 0.$$

By (6.3) it follows that each $k_{ij}^{\alpha} \in I_\alpha(E)$. q.e.d.

We now prove Theorem (*) of §2. Elements of $I_\alpha(E)$ are typically of the form

$$\omega = \sum_{i=0}^{q} \omega_i \cdot \theta_i , \quad \omega_i \in A^*(U_\alpha) .$$

Thus $I_\alpha(E)^{q+1} = 0$. If φ is an invariant polynomial of degree $> q$, then (6.4) and this remark imply

$$\varphi(k^\alpha) = 0 .$$

Thus $\varphi(k) = 0$ for $\deg(\varphi) > q$ and (*) follows immediately by our definition of the Pontryagin algebra.

As remarked in §2, this gives the global integrability theorem (*)′. By exactly parallel reasoning one also obtains the following global holomorphic analogue.

(*)′′ Theorem. If M is a complex analytic manifold, $T(M)$ the holomorphic tangent bundle, $E \subset T(M)$ a complex subbundle which is isomorphic to a holomorphic integrable subbundle $E' \subset T(M)$, $Q = T(M)/E$, and $q = \dim_{\mathbb{C}}(Q)$, then $\mathrm{Chern}^k(Q) = 0$ for $k > 2q$.

Using this theorem, we exhibit the first known counterexample to global integrability. Let CP^n denote the complex n-dimensional projective space, $T = T(CP^n)$ its holomorphic tangent bundle.

(6.5) Theorem. If n is odd, then T contains a holomorphic subbundle of complex codimension one. If, furthermore, $n > 1$, no holomorphic subbundle of T with codimension one is integrable.

Proof. Let $[n+1]$ denote the trivial bundle $CP^n \times C^{n+1}$ over CP^n. Each $x \in CP^n$ is a one dimensional subspace of C^{n+1}, hence one can define

$$S = \{(x,v) \in [n+1]: v \in x\}$$

$$Q = \{(x,v) \in [n+1]: v \perp x\} ,$$

holomorphic bundles of respective complex dimensions 1 and n, $[n+1] = S \oplus Q$. It is rather well known that there is a canonical isomorphism

$$T \cong \mathrm{Hom}(S,Q)$$

of holomorphic bundles. Indeed, given $\varphi \in \mathrm{Hom}(S_x,Q_x)$, $x \in CP^n$, choose any nonzero $(x,v) \in S_x$, and define

$$\sigma : U \to CP^n$$

on $U = \{z \in C : |z| < 1\}$, by letting $\sigma(z)$ be the one dimensional subspace of C^{n+1} containing $(1-z)v + z\varphi(v)$. Then σ is a holomorphic curve with $\sigma(0) = x$. The holomorphic tangent vector to CP^n at x determined by this curve will be labeled $\nu(\varphi)$. In this way one obtains the desired isomorphism

$$\nu : \mathrm{Hom}(S,Q) \to T(CP^n).$$

Let n be odd, choose a basis $v_1,\ldots,v_{n+1}$ of $(C^{n+1})^*$, and form the nondegenerate antisymmetric bilinear form

$$\omega = v_1 \wedge v_2 + v_3 \wedge v_4 + \cdots + v_n \wedge v_{n+1} .$$

Let H denote the dual bundle S^* and let H^2 be the symmetric square of H. Thus each $\eta \in H_x^2$ is simply a homogeneous function of degree 2 from $S_x \to C$. H^2 is a line bundle. Define

$$\omega_x : \mathrm{Hom}(S_x, Q_x) \to H_x^2$$

by

$$\omega_x(\varphi)(v) = \omega(v,\varphi(v)), \quad \forall\, v \in S_x .$$

Since ω is nondegenerate and $\varphi(v)$ ranges over the orthogonal complement of $v \neq 0$, ω_x is a surjection, $\forall\, x \in CP^n$, hence defines a holomorphic bundle surjection

$$\omega_* : T = \mathrm{Hom}(S,Q) \longrightarrow H^2 .$$

$\mathrm{Ker}(\omega_*)$ is a holomorphic subbundle of T of complex codimension one.

Finally, suppose that $n > 1$ and let $E \subset T$ be holomorphic of codimension one. If E is integrable,

$$\{c_1(T/E)\}^2 \in \mathrm{Chern}^4(T/E) = 0$$

by (*)''. Since $H^*(CP^n;C)$ is well known to be generated by 1 and by $\mu = c_1(S^*) \in H^2(CP^n;C)$ with the single relation $\mu^n = 0$, the above together with $n > 1$ implies that $c_1(T/E) = 0$. But

$$c(T) = c(E) \cdot c(T/E)$$

then implies $c_n(T) = 0$. On the other hand

$$T \oplus [1] = \mathrm{Hom}(S, Q \oplus S) = \mathrm{Hom}(S,[n+1]) = S^* \oplus \cdots \oplus S^* \text{ (n+1 times)},$$

so

$$c(T) = c(T \oplus [1]) = c(S^*)^{n+1} = (1 + \mu)^{n+1}$$

shows that $c_n(T) \neq 0$. This contradiction proves that E cannot be integrable. q.e.d.

As remarked in § 2, the proof of (*) has really proved something much stronger. We indicate an idea, due to Shulman, which exploits this stronger result.

Consider cohomology classes $[\alpha]$, $[\beta]$, $[\gamma]$ (in any suitable cohomology theory) such that

$$[\alpha]\cdot[\beta] = 0$$

$$[\beta]\cdot[\gamma] = 0 \quad .$$

Then, at the cocycle level,

$$\alpha\cdot\beta = \delta(x)$$

$$\beta\cdot\gamma = (-1)^{\deg(\alpha)}\delta(y)$$

so that $\alpha\cdot\beta\cdot\gamma$ is cohomologous to 0 in two different ways:

$$\alpha\cdot\beta\cdot\gamma = \delta(x\cdot\gamma)$$

$$\alpha\cdot\beta\cdot\gamma = \delta(\alpha\cdot y) \quad .$$

Thus $x\cdot\gamma - \alpha\cdot y$ is a cocycle and we set

$$< [\alpha], [\beta], [\gamma] > = [x\cdot\gamma - \alpha\cdot y] \in H^* \quad .$$

This is not well defined independently of the various choices, but it is easy to check that the indeterminacy is the ideal $I^*([\alpha],[\gamma])$ in H^* generated by $[\alpha]$ and $[\gamma]$. Thus $< [\alpha],[\beta],[\gamma] >$ is well defined as an element of $H^*/I^*([\alpha],[\gamma])$. This is called the Massey triple product of $[\alpha],[\beta],[\gamma]$. Higher order Massey products can also be defined [1].

The following is immediate from our definitions.

(6.6) Lemma. If $[\alpha]\cdot[\beta] = 0 = [\beta]\cdot[\gamma]$ and if the representative cocycles can so be chosen that $\alpha\cdot\beta = 0 = \beta\cdot\gamma$, then $< [\alpha],[\beta],[\gamma] > = 0$.

(6.7) Theorem. (Shulman) If E is integrable and $a,b,c \in \mathrm{Pont}^*(Q)$ are such that $\deg(a) + \deg(b) > 2q$ and $\deg(b) + \deg(c) > 2q$, then $< a,b,c >$ is defined and $= 0$.

Proof. Let k be the curvature of a basic connection and let $\varphi_1,\varphi_2,\varphi_3$ be invariant polynomials such that $a = [\varphi_1(k)]$, $b = [\varphi_2(k)]$, $c = [\varphi_3(k)]$. Then $a\cdot b = [(\varphi_1\varphi_2)(k)] = 0$ and $b\cdot c = [(\varphi_2\varphi_3)(k)] = 0$, so $< a,b,c >$ is defined. But actually,

$$\varphi_1(k) \cdot \varphi_2(k) = 0$$

$$\varphi_2(k) \cdot \varphi_3(k) = 0 ,$$

so by (6.6), $< a,b,c > = 0$. q.e.d.

For maximum usefulness of (6.7), it should be noted that the isomorphism between de Rham cohomology and Čech cohomology preserves Massey products. This follows directly from the fact that

$$\alpha : A^*(M) \longrightarrow K^*(U)$$
$$\beta : \check{C}^*(U;R) \longrightarrow K^*(U)$$

are ring homomorphisms.

We note also that similar treatments are possible for higher order Massey products. Using (6.7), Shulman has exhibited bundles which satisfy (*) but are not integrable.

References

1. W. S. Massey, Some higher order cohomology operations, Symposium Internacional de Topología Algebraica, Univ. Nac. Autonoma de Mexico and UNESCO, Mexico City, 1958.

7. Haefliger structures and the functor Γ_q. We assume here some acquaintance with the basic concepts of sheaves (cf. [1], [2]).

Over R^q construct the sheaf $\mathcal{D}$ of germs of local diffeomorphisms of $R^q \longrightarrow R^q$. That is, if $x \in R^q$, the stalk $\mathcal{D}_x$ is the set of germs at x of diffeomorphisms of open neighborhoods of x onto open sets of R^q. If $x,y \in R^q$ we adopt the notation $\mathcal{D}_{(x,y)}$ for the set $\{\gamma \in \mathcal{D}_x : \gamma(x) = y\}$.

(7.1) **Definition.** A Haefliger cocycle on a topological space X (also called a Γ_q-cocycle on X) consists of the following data:

1) An open cover $\{U_\alpha\}_{\alpha\in A}$ of X.

2) For each $\alpha \in A$, a continuous map $f_\alpha : U_\alpha \to R^q$ (the germ of which at $x \in U_\alpha$ will be denoted f_α^x).

3) For each $x \in U_\alpha \cap U_\beta$ a germ $\gamma_{\alpha\beta}^x \in \mathfrak{G}_{(f_\beta(x),f_\alpha(x))}$ such that

a) The assignment $x \to \gamma_{\alpha\beta}^x$ defines a continuous map

$$U_\alpha \cap U_\beta \to \mathfrak{G}$$

b) $f_\alpha^x = \gamma_{\alpha\beta}^x \circ f_\beta^x$

c) $\gamma_{\alpha\beta}^x \circ \gamma_{\beta\delta}^x = \gamma_{\alpha\delta}^x$.

(7.2) **Definition.** Γ_q-cocycles $c = \{U_\alpha, f_\alpha, \gamma^x{}_{\alpha\beta}\}_{\alpha,\beta\in A}$ and $c' = \{U_\lambda, f_\lambda, \gamma_{\lambda\mu}^x\}_{\lambda,\mu\in B}$ are called equivalent if $\exists$ cocycle corresponding to the covering $\{U_\zeta\}_{\zeta\in C}$, C = disjoint union of A and B, which restricts (in the obvious sense) to c on $\{U_\alpha\}_{\alpha\in A}$ and to c' on $\{U_\lambda\}_{\lambda\in B}$.

(7.3) **Exercise.** Show that the above relation of equivalence is an equivalence relation. [Hint. If $\{U_a, f_a, \gamma_{ab}^x\}_{a,b\in A\cup B}$ and $\{U_s, f_s, \gamma_{st}^x\}_{s,t\in B\cup C}$ restrict to the same Γ_q-cocycle on $\{U_\lambda\}_{\lambda\in B}$, then for $a\in A$, $s\in C$, $x \in U_a \cap U_s$, find $\lambda \in B$ with $x \in U_\lambda$ and try to define $\gamma_{sa}^x = \gamma_{s\lambda}^x \circ \gamma_{\lambda a}^x$. Show this independent of λ].

(7.4) **Definition.** A Haefliger structure (or Γ_q-structure) on X is an equivalence class of Haefliger cocycles on X. The set of all such structures on X is denoted $H^1(X;\Gamma_q)$.

Of course, the principal example of a Haefliger structure which we have in mind is a foliation. Indeed, an integrable subbundle $E \subset T(M)$ of codimension q allows us to cover M with coordinate patches U_α and to find submersions

$$f_\alpha : U_\alpha \to R^q$$

such that $E|U_\alpha = \mathrm{Ker}(df_\alpha)$. The fact that the maps f_α are submersions together with the fact that $\mathrm{Ker}(df_\alpha)$ and $\mathrm{Ker}(df_\beta)$ agree on $U_\alpha \cap U_\beta$ implies that for each $x \in U_\alpha \cap U_\beta$ there is a unique $\gamma^x_{\alpha\beta} \in \mathcal{S}_{(f_\beta(x), f_\alpha(x))}$ such that $\gamma^x_{\alpha\beta} \circ f^x_\beta = f^x_\alpha$. This uniqueness makes it easy to verify the various properties of a Haefliger cocycle. Finally, two cocycles obtained from the same foliation are clearly equivalent, hence the foliation defines a unique element of $H^1(M;\Gamma_q)$.

One defines the normal bundle Q of a Haefliger structure by the GL_q-cocycle $g_{\alpha\beta}(x) = d(\gamma^x_{\alpha\beta})$. Equivalent Γ_q-cocycles give equivalent GL_q-cocycles, hence the normal bundle of a Γ_q structure is uniquely determined up to isomorphism. For the case of a foliation E, Q is just $T(M)/E$.

Remark that a continuous map $g : Y \to X$ can be used to pull back a Γ_q-cocycle on X to one on Y. Indeed, $\{U_\alpha, f_\alpha, \gamma^x_{\alpha\beta}\}_{\alpha,\beta \in A}$ pulls back to $\{\bar{U}_\alpha, \bar{f}_\alpha, \bar{\gamma}^y_{\alpha\beta}\}_{\alpha,\beta \in A}$ where

$$\bar{U}_\alpha = g^{-1}(U_\alpha)$$

$$\bar{f}_\alpha = f_\alpha \circ g$$

$$\bar{\gamma}^y_{\alpha\beta} = \gamma^{g(y)}_{\alpha\beta} .$$

This pull-back by g respects equivalence, hence defines

$$g^* : H^1(X;\Gamma_q) \to H^1(Y;\Gamma_q).$$

This makes $H^1(;\Gamma_q)$ into a contravariant functor from the category of topological spaces to the category of sets.

It is an unpleasant fact of life that this functor is not homotopy invariant. As an example, consider the maps

$$g_0 : R^q \to R^q, \quad g_0(x) = 0, \quad \forall\, x \in R^q$$

$$g_1 : R^q \to R^q, \quad g_1(x) = x, \quad \forall\, x \in R^q .$$

These maps are homotopic. On the image space R^q consider the Γ_q-cocycle c with just one element $U_1 = R^q$ in the open cover, just one $f_1 = id : U_1 \to R^q$, and every γ^x_{11} = germ of the identity diffeomorphism. We claim that $g_0^*(c)$ is not equivalent to $g_1^*(c)$. To see this clearly we introduce a definition and a lemma.

(7.5) Definition. If $c = \{U_\alpha, f_\alpha, \gamma^x_{\alpha\beta}\}_{\alpha,\beta \in A}$ is a Γ_q-cocycle on X, the level set of c through $x \in U_\alpha$ is defined to be $X^x_\alpha = \{y \in U_\alpha : f_\alpha(y) = f_\alpha(x)\}$.

(7.6) Lemma. If $\{U_\alpha, f_\alpha, \gamma^x_{\alpha\beta}\}_{\alpha,\beta \in A}$ and $\{U_\lambda, f_\lambda, \gamma^x_{\lambda\mu}\}_{\lambda,\mu \in B}$ are equivalent Γ_q-cocyles on X, then for any $x \in U_\alpha \cap U_\lambda$, $\alpha \in A$, $\lambda \in B$, there is an open neighborhood W of x such that $X^x_\alpha \cap W = X^x_\lambda \cap W$.

The proof of (7.6) is an elementary application of definitions. Returning to our example, $g_0^*(c)$ and $g_1^*(c)$ cannot be equivalent since the only level set for $g_0^*(c)$ is R^q while the level sets for $g_1^*(c)$ are the single points of R^q.

In order to obtain a homotopy invariant functor, we impose a further equivalence relation on $H^1(X;\Gamma_q)$.

(7.7) Definition. If $\alpha, \alpha' \in H^1(X;\Gamma_q)$ we say that α and α' are homotopic and write $\alpha \simeq \alpha'$ if and only if $\exists\ \lambda \in H^1(X \times I; \Gamma_q)$ such that $\alpha = i_0^*(\lambda)$, $\alpha' = i_1^*(\lambda)$.

Here, of course, $i_0, i_1 : X \to X \times I$ are the usual face maps.

(7.8) Exercises. 1) Prove that homotopy is an equivalence relation on $H^1(X;\Gamma_q)$.

2) If $f : X \to Y$ is continuous, prove that $f^* : H^1(Y;\Gamma_q) \to H^1(X;\Gamma_q)$ preserves the relation of homotopy.

3) If $f, g : X \to Y$ are homotopic maps, $\alpha \in H^1(Y;\Gamma_q)$, then $f^*(\alpha) \simeq g^*(\alpha)$.

(7.9) Definition. $\Gamma_q(X)$ = set of homotopy classes in $H^1(X;\Gamma_q)$.

By 2) and 3) of (7.8) we see that $\Gamma_q(\ )$ is a homotopy invariant contravariant functor.

Haefliger [4], [5] has shown that the functor Γ_q is representable. That is, one can construct a space $B\Gamma_q$ such that, for "reasonable" spaces X, there is a canonical one-one correspondence between the set $\Gamma_q(X)$ and the set of homotopy classes of maps $[X,B\Gamma_q]$. Furthermore, this correspondence is natural with respect to the set maps induced by continuous maps, so we have an equivalence of functors

$$\Gamma_q(\ ) \cong [\ ,B\Gamma_q]$$

on a reasonable category of spaces. There are three ways of showing this.

1) Γ_q is shown to be representable by checking the axioms of Brown [2] for representable functors. This works for CW-complexes X.

2) One can mimic Milnor's construction of the classifying space of a topological group [6]. This works for all paracompact spaces X.

3) One can mimic the "abstract nonsense" approach to classifying spaces due to Graeme Segal [7].

The first approach is written up in [5] and the second in [4]. We will take a look at the third approach in the next section.

If M is a smooth manifold, a smooth Γ_q-cocycle is defined in the obvious way, and an element of $H^1(M;\Gamma_q)$ is said to be a smooth Haefliger structure if it is represented by a smooth Γ_q-cocycle. One has generalizations of the theorems of § 6 to these smooth structures.

(7.10) Theorem. If Q is the normal bundle to a smooth Γ_q-structure γ on M, then $\mathrm{Pont}^k(Q) = 0$ for $k > 2q$. If $a, b, c \in \mathrm{Pont}^*(Q)$ with $\deg(a) + \deg(b) > 2q$, $\deg(b) + \deg(c) > 2q$, then $\langle a,b,c \rangle = 0$.

The proof proceeds by analogy with the proofs in § 6. Over $U_\alpha \subset M$ let ∇^α be the connection obtained by pulling back the standard connection on R^q by f_α. Patch these together with a partition of unity and prove that the curvature matrices k^α for the resulting ∇ have $k^\alpha_{ij} \in I_\alpha(\gamma)$, the ideal in $A^*(U_\alpha)$ generated by those 1-forms which locally are pull-backs via f_α of 1-forms on R^q. Clearly $I_\alpha(\gamma)^{q+1} = 0$, so $\varphi(k) = 0$ for any invariant φ with $\deg(\varphi) > q$.

In §10, we generalize $\mathrm{Pont}^k(Q) = 0$, $k > 2q$, to the normal bundle of any Γ_q-structure on any reasonable space X.

References

1. G. Bredon, Sheaf Theory, McGraw-Hill, New York, 1967.

2. E. H. Brown, Abstract homotopy theory, Trans. Amer. Math. Soc. 119 (1965), pp. 79-85.

3. R. Godement, Topologie Algebrique et Theoric des Faisceaux, Hermann, Paris, 1958.

4. A. Haefliger, Homotopy and integrability, Lecture Notes in Mathematics, No. 197, Springer-Verlag, New York, 1971, pp. 133-163.

5. A. Haefliger, Feuilletages sur les variétés ouvertes, Topology 9 (1970), pp. 183-194.

6. D. Husemoller, Fibre Bundles, McGraw-Hill, New York, 1966.

7. G. Segal, Classifying spaces and spectral sequences, Institut des Hautes Etudes Scientifiques, Publications Mathematiques, No. 34 (1968), pp. 105-112.

8. Topological categories and classifying spaces.

The basic reference for this section is [1].

Recall the notion of a category C. We write $C = \mathfrak{O} \cup \mathfrak{M}$ where $\mathfrak{O}$ is the class of objects of C and $\mathfrak{M}$ is the class of morphisms of C described as follows. For each pair $X, Y \in \mathfrak{O}$, Hom (X,Y) is a set and

$$\mathfrak{M} = \coprod_{(X,Y) \in \mathfrak{O} \times \mathfrak{O}} \mathrm{Hom}(X,Y),$$

a disjoint union. The basic axioms are:

1) If $X,Y,Z \in \mathfrak{O}$, there is a map (called composition)

$$\mathrm{Hom}(Y,Z) \times \mathrm{Hom}(X,Y) \to \mathrm{Hom}(X,Z)$$

(written $(f,g) \mapsto f \circ g$).

2) Composition is associative.

3) For each $X \in \mathfrak{O}$, $\exists\, 1_X \in \mathrm{Hom}(X,X)$ such that $f \circ 1_X = f$ and $1_X \circ g = g$ whenever these compositions are defined.

Typical examples are the category of sets and set mappings, the category of topological spaces and continuous mappings, the category of groups and homomorphisms, etc. Not all categories, however, need be so enormous. Indeed, in order to avoid logical difficulties in what follows, we consider only "small" categories (ones in which $\mathfrak{O}$ and $\mathfrak{m}$ are sets).

As an example of a small category, let G be a group and define the category whose only object is G with Hom(G,G) = G, and define composition to be the group operation.

As another example, let X be a topological space, $\{U_\alpha\}_{\alpha \in A}$ an open cover of X. Define a category whose objects are the finite subsets $\Sigma \subset A$ such that

$$U_\Sigma = \bigcap_{\alpha \in \Sigma} U_\alpha \neq \phi$$

and let

$$\mathrm{Hom}(\Sigma,\Sigma') = \begin{cases} \text{inclusion map } U_\Sigma \hookrightarrow U_{\Sigma'} \text{ if defined} \\ \phi \quad \text{otherwise} \end{cases} .$$

Composition is composition in the usual sense.

(8.1) Definition. A topological category $C = \mathfrak{O} \cup \mathfrak{m}$ is a small category such that $\mathfrak{O}$ and $\mathfrak{m}$ are topological spaces and the following maps are continuous.

1) $\mathfrak{m} \to \mathfrak{O} \times \mathfrak{O}$ defined by $f \mapsto (X,Y)$, $\forall f \in \mathrm{Hom}(X,Y)$.

2) Composition $\mathfrak{m} * \mathfrak{m} \to \mathfrak{m}$, where $\mathfrak{m} * \mathfrak{m} = \{(f,g) \in \mathfrak{m} \times \mathfrak{m} : f \circ g$ is defined$\}$.

3) $\mathfrak{O} \to \mathfrak{m}$ defined by $X \mapsto 1_X$.

Remark that any small category can be regarded as a topological category with the discrete topology on $\mathfrak{O}$ and $\mathfrak{m}$.

As a first example, let G be a topological group and again form the category with one object G and Hom(G,G) = G. This clearly satisfies (8.1).

As a slightly less trivial example, let X be a topological space,

$U = \{U_\alpha\}_{\alpha \in A}$ an open cover of X, and form the topological category X_U whose objects are the pairs (Σ, x) where $\Sigma \subset A$ is as in our earlier example and $x \in U_\Sigma$. $\mathrm{Hom}((\Sigma,x),(\Sigma', x'))$ is empty if either $x \neq x'$, or $U_\Sigma \not\subset U_{\Sigma'}$. Otherwise it consists of the single inclusion $(U_\Sigma, x) \to (U_{\Sigma'}, x)$. Topologize $\mathfrak{O}$ by taking as basic neighborhoods sets of the form $\{(\Sigma,x) : \Sigma$ fixed, $x \in W \subset U_\Sigma$, W open in $U_\Sigma\}$. Similarly, topologize $\mathfrak{m}$ by basic neighborhoods of the form $\{(U_\Sigma,x) \hookrightarrow (U_{\Sigma'}, x) : \Sigma$ and Σ' fixed, $x \in W \subset U_\Sigma$, W open in $U_\Sigma\}$.

As a third example, we construct the topological category Γ_q which is pertinent to the study of Haefliger structures. Let $\mathfrak{O} = R^q$ with the usual topology. Given $x, y \in R^q$, let $\mathrm{Hom}(x,y) = \mathcal{D}_{(x,y)}$ as in § 7. Thus $\mathfrak{m} = \mathcal{D}$ with the sheaf topology. It is easy to check that composition $\mathcal{D} * \mathcal{D} \to \mathcal{D}$ is continuous. The map $x \to 1_x$ (the germ at x of the identity diffeomorphism) defines a natural continuous section $R^q \to \mathcal{D}$. Finally, let

$$\iota : \mathcal{D} \to \mathcal{D}$$

be the homeomorphism which to each germ γ of a local diffeomorphism g assigns the germ γ^{-1} of g^{-1}. If

$$\pi : \mathcal{D} \to R^q$$

is the sheaf projection, then the map $\mathfrak{m} \to \mathfrak{O} \times \mathfrak{O}$ in (8.1) becomes

$$\mathcal{D} \to R^q \times R^q$$

$$\gamma \to (\pi(\gamma), \pi \circ \iota(\gamma)) \quad ,$$

hence is continuous.

We next take up the construction of the classifying space of a topological category $C = \mathfrak{O} \cup \mathfrak{m}$. Let $A_0C = \mathfrak{O}$ and, if $n > 0$ is an integer, let A_nC denote the set of all finite sequences

$$X_0 \xrightarrow[f_1]{} X_1 \xrightarrow[f_2]{} \cdots \xrightarrow[f_n]{} X_n$$

where $X_i \in \Theta$ and $f_i \in \mathrm{Hom}(X_{i-1}, X_i)$. A_0C has the topology of Θ and A_nC is topologized as a subset of $\mathfrak{M} \times \mathfrak{M} \times \cdots \times \mathfrak{M}$. Define functions

$$\partial_i : A_nC \to A_{n-1}C , \quad 0 \leq i \leq n$$

$$s_i : A_nC \to A_{n+1}C , \quad 0 \leq i \leq n$$

by

$$\partial_0(X_0 \xrightarrow[f_1]{} X_1 \xrightarrow[f_2]{} \cdots \xrightarrow[f_n]{} X_n) = X_1 \xrightarrow[f_2]{} \cdots \xrightarrow[f_n]{} X_n$$

$$\partial_n(X_0 \xrightarrow[f_1]{} X_1 \xrightarrow[f_2]{} \cdots \xrightarrow[f_n]{} X_n) = X_0 \xrightarrow[f_1]{} \cdots \xrightarrow[f_{n-1}]{} X_{n-1}$$

$$\partial_i(X_0 \xrightarrow[f_1]{} \cdots \xrightarrow[f_n]{} X_n) = X_0 \xrightarrow[f_1]{} \cdots \to X_{i-1} \xrightarrow[f_{i+1} \circ f_i]{} X_{i+1} \to \cdots \xrightarrow[f_n]{} X_n$$

$$s_i(X_0 \xrightarrow[f_1]{} \cdots \xrightarrow[f_n]{} X_n) = X_0 \xrightarrow[f_1]{} \cdots \xrightarrow[f_i]{} X_i \xrightarrow[1_{X_i}]{} X_i \xrightarrow[f_{i+1}]{} \cdots \xrightarrow[f_n]{} X_n .$$

Let Δ_n denote the standard n-simplex with its usual topology, $E_0, \ldots, E_n$ the vertices, and define functions

$$\epsilon_i : \Delta_{n-1} \to \Delta_n , \quad 0 \leq i \leq n$$

$$p_i : \Delta_{n+1} \to \Delta_n , \quad 0 \leq i \leq n$$

by the linear extensions of

$$\epsilon_i(E_j) = \begin{cases} E_j , & 0 \leq j \leq i-1 \\ E_{j+1}, & i \leq j \leq n-1 \end{cases}$$

$$p_i(E_j) = \begin{cases} E_j , & 0 \leq j \leq i \\ E_{j-1}, & i+1 \leq j \leq n+1 . \end{cases}$$

Then on the disjoint union of the topological spaces $A_nC \times \Delta_n$, $0 \leq n < \infty$, make the identifications

$$(a, \epsilon_i(t)) \sim (\partial_i(a), t)$$

$$(s_i(a), t) \sim (a, p_i(t)).$$

The resulting quotient space is designated by BC and is called the classifying space of C.

(8.2) Definition. Let $C = \mathfrak{O} \cup \mathfrak{M}$ and $C' = \mathfrak{O}' \cup \mathfrak{M}'$ be categories. A function $F : C \to C'$ is called a (covariant) functor if

(1) $F(\mathfrak{O}) \subset \mathfrak{O}'$, $F(\mathfrak{M}) \subset \mathfrak{M}'$

(2) $\varphi \in \mathrm{Hom}(X,Y) \Rightarrow F(\varphi) \in \mathrm{Hom}(F(X),F(Y))$.

(3) $F(\varphi \circ \psi) = F(\varphi) \circ F(\psi)$ whenever the compositions are defined.

(4) $F(1_X) = 1_{F(X)}$, $\forall X \in \mathfrak{O}$.

If C and C' are topological categories, a continuous functor $F : C \to C'$ is a functor such that $F : \mathfrak{O} \to \mathfrak{O}'$ and $F : \mathfrak{M} \to \mathfrak{M}'$ are both continuous.

Clearly a continuous functor $F : C \to C'$ induces a continuous map

$$BF : BC \to BC'$$

and

$$B(F \circ G) = BF \circ BG$$

$$B(1_C) = 1_{BC}$$

where $1_C : C \to C$ is the identity functor. B is a functor from the category of topological categories and continuous functors to the category of topological spaces and continuous maps.

We consider two important examples. Let X be a space, $U = \{U_\alpha\}_{\alpha \in A}$ an open cover of X, and form the topological category X_U defined above. The topological group $GL_q = GL(q,R)$ also defines a topological category as described earlier, and we consider a continuous functor

$$F : X_U \to GL_q .$$

For $\alpha, \beta \in A$, $x \in U_\alpha \cap U_\beta$, there are unique morphisms

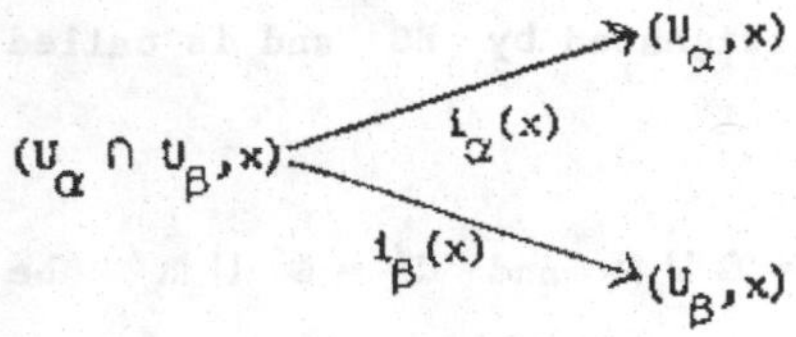

in X_U. Define

$$g_{\alpha\beta} : U_\alpha \cap U_\beta \longrightarrow GL_q$$

by

$$(8.3) \qquad g_{\alpha\beta}(x) = F(i_\alpha(x)) \cdot F(i_\beta(x))^{-1} .$$

By the continuity of F, $g_{\alpha\beta}$ is continuous.

(8.4) **Lemma.** $\{g_{\alpha\beta}\}_{\alpha,\beta\in A}$ satisfies the cocycle condition (2.2).

Proof. If $U_\alpha \cap U_\beta \cap U_\gamma \neq \emptyset$, then for any $x \in U_\alpha \cap U_\beta \cap U_\gamma$ we have the commutative diagram

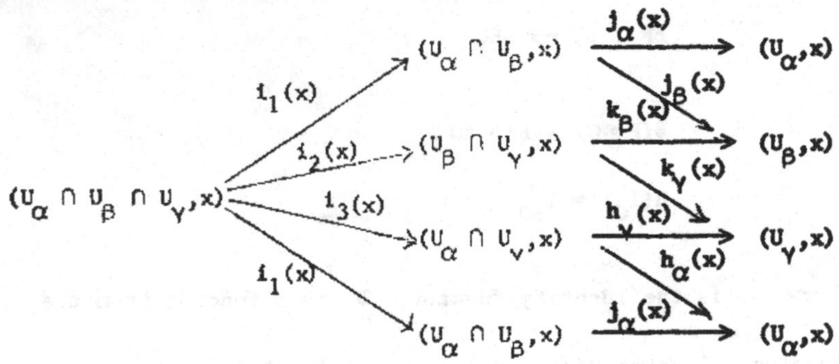

$g_{\alpha\beta}(x) = F(j_\alpha(x)) \cdot F(j_\beta(x))^{-1} = F(j_\alpha(x)) \cdot F(i_1(x)) \cdot F(i_1(x))^{-1} \cdot F(j_\beta(x))^{-1}$

$= F(j_\alpha(x) \circ i_1(x)) \cdot F(j_\beta(x) \circ i_1(x))^{-1}$. Similarly,

$g_{\beta\gamma}(x) = F(k_\beta(x) \circ i_2(x)) \cdot F(k_\gamma(x) \circ i_2(x))^{-1} = F(j_\beta(x) \circ i_1(x)) \cdot F(k_\gamma(x) \circ i_2(x))^{-1}$.

Thus $\quad g_{\alpha\beta}(x) g_{\beta\gamma}(x) = F(j_\alpha(x) \circ i_1(x)) \cdot F(k_\gamma(x) \circ i_2(x))^{-1}$

$= F(h_\alpha(x) \circ i_3(x)) \cdot F(h_\gamma(x) \circ i_3(x))^{-1} = F(h_\alpha(x)) \cdot F(h_\gamma(x))^{-1} = g_{\alpha\gamma}(x)$. q.e.d.

Conversely, let a GL_q-cocycle $\{g_{\alpha\beta}\}_{\alpha,\beta\in A}$ be given. Extend this to a cocycle $\{g_{\Sigma\Sigma'}\}$ over $\{U_\Sigma\}$. This can be done, e.g., by linearly ordering A and defining $g_{\Sigma\Sigma'}(x) = g_{\alpha\alpha'}(x)$ where $\alpha \in \Sigma$ and $\alpha' \in \Sigma'$ are the maximal elements. Then define

$$F : X_U \to GL_q$$

by

$$F((U_\Sigma,x) \to (U_{\Sigma'},x)) = g_{\Sigma'\Sigma}(x).$$

This is continuous and

$$F(1_{(U_\Sigma,x)}) = 1_{GL_q}$$

$$F([(U_{\Sigma'},x) \to (U_{\Sigma''},x)] \circ [(U_\Sigma,x) \to (U_{\Sigma'}(x))])$$

$$= F((U_\Sigma,x) \to (U_{\Sigma''},x)) = g_{\Sigma''\Sigma}(x)$$

$$= g_{\Sigma''\Sigma'}(x)g_{\Sigma'\Sigma}(x) = F((U_{\Sigma'},x) \to (U_{\Sigma''},x))F((U_\Sigma,x) \to (U_{\Sigma'},x)) \quad .$$

Thus F is a continuous functor. Indeed, F defined in this way is a functor if and only if $\{g_{\alpha\beta}\}$ satisfies the cocycle condition (2.2). Furthermore, it is trivial to see that the cocycle constructed from F by (8.3) is the original cocycle $\{g_{\alpha\beta}\}$. To complete the picture, we need appropriate notions of equivalence.

It is well known (and an easy exercise to prove) that two cocycles $\{g_{\alpha\beta}\}$ and $\{g'_{\alpha\beta}\}$ defined on the same open cover $\{U_\alpha\}_{\alpha\in A}$ come from isomorphic vector bundles if and only if there is a collection

$$\{\theta_\alpha : U_\alpha \to GL_q\}_{\alpha\in A}$$

of continuous functions such that

$$g'_{\alpha\beta}(x) = \theta_\alpha(x)g_{\alpha\beta}(x)\theta_\beta(x)^{-1} ,$$

$\forall\, x \in U_\alpha \cap U_\beta$, $\forall\, \alpha,\beta \in A$. We call such cocycles equivalent (or cohomologous) and denote the set of equivalence classes by $H^1(U;GL_q)$.

(8.5) Definition. If $F, H : C \to C'$ are continuous functors, a natural transformation $\theta : F \to H$ is a continuous map

$$\theta : \mathfrak{O} \to \mathfrak{m}'$$

(where $\mathfrak{O}$ denotes the set of objects in C and $\mathfrak{m}'$ the set of morphisms in C') such that

$$\theta(L) \in \mathrm{Hom}(F(L),H(L)),\ \forall\, L \in \mathfrak{O} \quad ,$$

and for any $\alpha \in \mathrm{Hom}(L,L')$, the diagram

$$\begin{array}{ccc} F(L) & \xrightarrow{F(\alpha)} & F(L') \\ \downarrow{\scriptstyle \theta(L)} & & \downarrow{\scriptstyle \theta(L')} \\ H(L) & \xrightarrow{H(\alpha)} & H(L') \end{array}$$

is commutative, $\forall\, L,L' \in \mathfrak{O}$. F and H are said to be isomorphic (written $F \cong H$) if there are natural transformations

$$\theta : F \to H$$
$$\varphi : H \to F$$

such that $\theta(L) \circ \varphi(L) = 1_{H(L)}$ and $\varphi(L) \circ \theta(L) = 1_{F(L)}$, $\forall\, L \in \mathfrak{O}$. The set of isomorphism classes of continuous functors $C \to C'$ will be denoted $\pi[C,C']$.

(8.6) Theorem. The construction (8.3) sets up a one-one correspondence between $H^1(U;GL_q)$ and $\pi[X_U,GL_q]$.

Proof. Given $F_1, F_2 : X_U \to GL_q$, suppose $\theta : F_2 \to F_1$ is an isomorphism. Let $\{g^1_{\alpha\beta}\}$ and $\{g^2_{\alpha\beta}\}$ be the cocycles defined via (8.3) by F_1 and F_2 respectively. Write $\theta_\alpha(x) = \theta(U_\alpha,x)$. Then, for $x \in U_\alpha \cap U_\beta$,

$$\begin{aligned} g^1_{\alpha\beta}(x) &= F_1(i_\alpha(x))F_1(i_\beta(x))^{-1} \\ &= \theta_\alpha(x)F_2(i_\alpha(x))\theta(U_\alpha \cap U_\beta,x)^{-1}\theta(U_\alpha \cap U_\beta,x)F_2(i_\beta(x))^{-1}\theta_\beta(x)^{-1} \\ &= \theta_\alpha(x)g^2_{\alpha\beta}(x)\theta_\beta(x)^{-1} \ . \end{aligned}$$

Thus $\pi[X_U,GL_q]$ is mapped in a well defined way into $H^1(U;GL_q)$.

If F_1 and F_2 give rise to equivalent cocycles $g^1_{\alpha\beta} = \theta_\alpha g^2_{\alpha\beta}\theta^{-1}_\beta$, then, as in the remarks following the proof of (8.4), extend them to cocycles $\{g^1_{\Sigma\Sigma'}\}$ and $\{g^2_{\Sigma\Sigma'}\}$ over the cover $\{U_\Sigma\}_{\Sigma \text{ finite} \subset A}$, and note that these cocycles are still equivalent, hence obtain $g^1_{\Sigma\Sigma'} = \theta_\Sigma g^2_{\Sigma\Sigma'} \theta^{-1}_{\Sigma'}$. Define

$$\theta(U_\Sigma,x) : F_2(U_\Sigma,x) \to F_1(U_\Sigma,x)$$

by

$$\theta(U_\Sigma,x) = \theta_\Sigma(x) \quad .$$

This is readily seen to be an isomorphism of continuous functors, so $\pi[X_U,GL_q] \to H^1(U;GL_q)$ is one-one. It is onto by the remarks following the proof of (8.4). q.e.d.

Thus, $\pi[X_U,GL_q]$ is in natural one-one correspondence with the set of isomorphism classes of q-dimensional vector bundles over X for which each $U_\alpha \in U$ is a trivializing neighborhood. Thus, if each U_α is contractible then $\pi[X_U,GL_q]$ is identified with the set of isomorphism classes of q-dimensional vector bundles over X.

For numerable coverings U (i.e., open coverings which admit a subordinate partition of unity) the space BX_U has canonically the homotopy type of X [1, Prop. (4.1)]. Any continuous functor $F : X_U \to GL_q$ defines a continuous map $BF : BX_U \to BGL_q$, hence defines a unique homotopy class

$$\widetilde{BF} \in [X,BGL_q] \ .$$

By [1, Prop. (2.1)], $F_1 \cong F_2$ implies $\widetilde{BF_1} = \widetilde{BF_2}$. Thus BGL_q appears to be a good candidate up to homotopy for the classifying space for vector bundles described in a quite different way in §2. [1,p.107] sketches the relation between this construction and the Milnor construction.

With the above discussion as motivation, consider the topological category Γ_q and form the space $B\Gamma_q$. Mimicry of the above discussion with suitable adjustments shows that giving a Haefliger cocycle amounts to giving a continuous functor

$$F : X_U \to \Gamma_q ,$$

hence (again assuming U numerable) the Haefliger cocycle will determine an element

$$\widetilde{BF} \in [X, B\Gamma_q].$$

Thus $B\Gamma_q$ is an excellent candidate for the Γ_q-classifying space discussed in § 7. We will not attempt a more detailed treatment here.

Remark. There is a continuous functor

$$\nu : \Gamma_q \to GL_q$$

defined by

$$\nu(\gamma^x) = d\gamma^x ,$$

the Jacobian at x of any local diffeomorphism whose germ is $\gamma^x \in \mathrm{Hom}(x,y)$. This gives rise to a very important continuous map

$$B\nu : B\Gamma_q \to BGL_q .$$

Indeed, in any homotopy commutative triangle

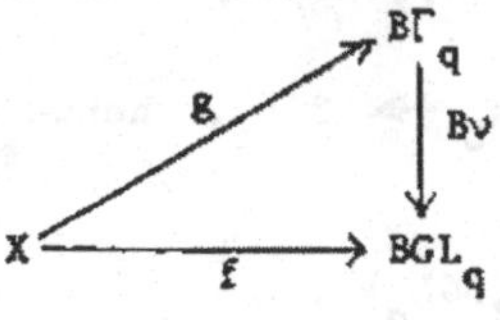

g is the classifying map of an element of $\Gamma_q(X)$ whose normal bundle is classified by f.

References

1. G. Segal, Classifying spaces and spectral sequences, Institut des Hautes Études Scientifiques, Publications Mathematiques, No. 34 (1968), pp. 105-112.

9. Integrable homotopy and $Fol_q(M)$. For M a manifold, we define an equivalence relation called "integrable homotopy" between codimension q foliations of M and so obtain the set $Fol_q(M)$ of integrable homotopy classes. If M is an open manifold, an application due to Haefliger [1] of the Phillips-Gromov theorem will show how to describe $Fol_q(M)$ by means of the space $B\Gamma_q$.

(9.1) Definition. Let E_0 and E_1 be foliations of codimension q on M (i.e., E_0 and E_1 are integrable subbundles of T(M) of codimension q). An integrable homotopy between these foliations is a subbundle $\tilde{E} \subset T(M \times R)$ of codimension q such that

1) $\tilde{E}$ is integrable.

2) $\tilde{E}_{(x,t)} + T_{(x,t)}(M \times \{t\}) = T_{(x,t)}(M \times R)$

(i.e., $\tilde{E}$ is transversal to each slice $M \times \{t\}$).

3) $\tilde{E}_{(x,0)} \cap T_{(x,0)}(M \times \{0\}) = E_{0x}$ and $\tilde{E}_{(x,1)} \cap T_{(x,1)}(M \times 1) = E_{1x}$.

By standard techniques this is shown to be an equivalence relation, so we obtain the set $Fol_q(M)$ as desired.

If M is n-dimensional, the classifying map

$$g_M : M \to BGL_n$$

for the tangent bundle is called the Gauss map. If M has a codimension q foliation E, the tangent bundle splits

$$T(M) = Q \oplus E$$

and so there is a homotopy lift of the Gauss map

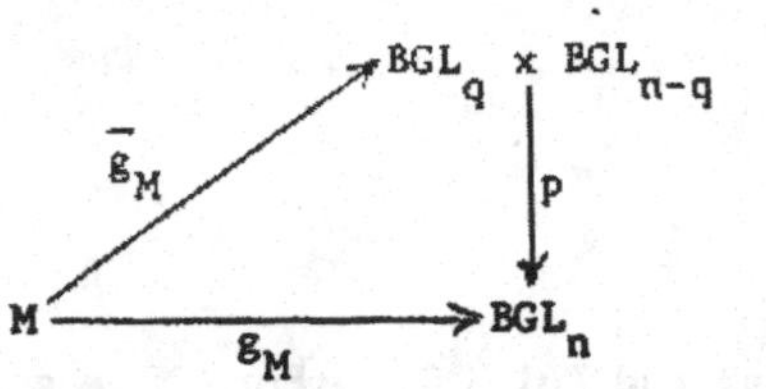

.

Since Q is the normal bundle of a foliation, $\bar{g}_M$ admits a homotopy lift

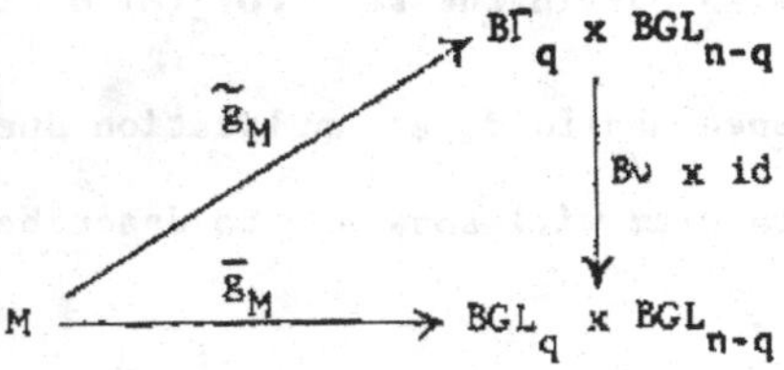

.

Set $\rho = p \circ (B\nu \times id)$.

(9.2) Theorem. (Phillips, Gromov, Haefliger). If M is an open manifold (i.e., $\partial M = \emptyset$ and M has no compact component) the above construction sets up a one-one correspondence between $Fol_q(M)$ and the set of homotopy classes of homotopy lifts $\tilde{g}_M$ of g_M :

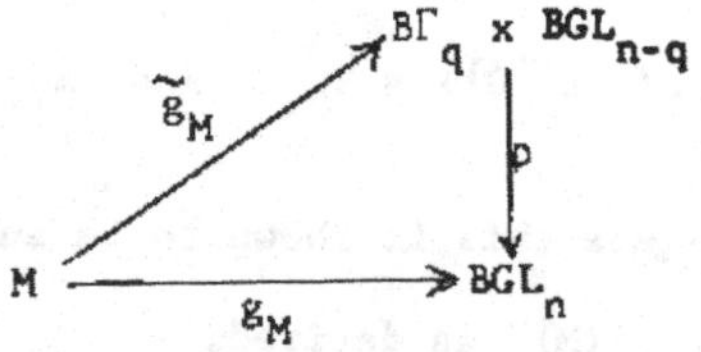

.

The proof of this theorem is quite difficult. It depends on the Phillips-Gromov theorem, of which an account by Poenaru will be found in [2]. The theorem in the form given here is then established by Haefliger in [1].

References

1. A. Haefliger, Feuilletages sur les variétés ouvertes, Topology 9 (1970), pp. 183-194.

2. V. Poenaru, Homotopy theory and differentiable singularities, Lecture Notes in Mathematics, No. 197, Springer-Verlag, N. Y., 1971, pp. 106-132.

10. The topology of $B\Gamma_q$. In this section we sketch in fair detail some results on the algebraic topological invariants of $B\Gamma_q$. The most novel of these results will be the construction of certain "exotic" characteristic classes in $H^*(B\Gamma_q;\mathbb{C})$, but first we discuss some facts about the map $B\nu$ which by now are standard in the theory of Γ_q-structures. The first of these is an elementary consequence of (*).

(10.1) Theorem. $B\nu^* : H^k(BGL_q;\mathbb{R}) \to H^k(B\Gamma_q;\mathbb{R})$ is zero if $k > 2q$.

Proof. We use singular cohomology. If the assertion is false, choose $\psi \in H^k(BGL_q;\mathbb{R})$, $k > 2q$, such that $\varphi = B\nu^*(\psi) \neq 0$. Then there is a homology class $\sigma \in H_k(BGL_q;\mathbb{R})$ such that $\varphi(\sigma) \neq 0$. By the definition of singular homology, there is a finite polyhedron P, a homology class $\sigma' \in H_k(P;\mathbb{R})$, and a continuous map

$$s : P \to B\Gamma_q$$

(the "geometric realization" of σ) such that $s_*(\sigma') = \sigma$. s determines a homotopy class of Γ_q-structures on P and by [1, p.188] we can thicken P to an open manifold M having the same homotopy type as P such that

$$s : M \to B\Gamma_q$$

corresponds to a foliation of M. Furthermore, by construction

$s^*(\varphi) \neq 0$ in $H^k(M;k)$. Let

$$f = B\nu \circ s : M \to BGL_q ,$$

Q the corresponding bundle on M. Q is the normal bundle to our foliation of M, hence $\mathrm{Pont}^k(Q) = 0$. But

$$0 \neq s^*(\varphi) = s^* B\nu^*(\psi) = f^*(\psi) \in \mathrm{Pont}^k(Q) ,$$

a contradiction. q.e.d.

(10.2) Corollary. Let $\gamma \in \Gamma_q(X)$, Q the normal bundle of γ. Then $\mathrm{Pont}^k(Q) = 0$, $k > 2q$.

This is the generalization of (*) promised at the end of §7. It is an immediate corollary of (10.1).

As is well known, from the point of view of homotopy theory, every continuous map is a fibration. Thus let $F\Gamma_q$ denote the homotopy theoretic fiber of

$$B\nu : B\Gamma_q \to BGL_q .$$

We sketch the proof of the following theorem from [1].

(10.3) Theorem. $\pi_i(F\Gamma_q) = 0$, $0 \leq i \leq q$. Consequently, $B\nu^* : \pi_i(B\Gamma_q) \to \pi_i(BGL_q)$ is an isomorphism, $0 \leq i \leq q$, and a surjection for $i = q + 1$.

If $i < q$, (10.3) is deduced from (9.2) as follows. $S^i \times R^{q-i}$ is parallelizable, hence for the Gauss map

$$g : S^i \times R^{q-i} \to BGL_q$$

we may take a constant map. Then by (9.2) the homotopy classes of lifts

$$\tilde{g} : S^i \times R^{q-i} \to B\Gamma_q$$

correspond one-one to $\mathrm{Fol}_q(S^i \times R^{q-i})$ and furthermore, g being constant,

$\tilde{g}$ factors through $F\Gamma_q \subset B\Gamma_q$ as do all covering homotopies. This says that

$$\pi_i(F\Gamma_q) = \mathrm{Fol}_q(S^i \times R^{q-i}) ,$$

$0 \leq i < q$. But for any q-dimensional manifold M, it is clear that $\mathrm{Fol}_q(M)$ has only one element, so $\pi_i(F\Gamma_q) = 0$, $0 \leq i < q$.

For the case $i = q$, the proof proceeds differently. One must consider any continuous map $f : S^q \to F\Gamma_q$ and the element $\gamma \in \Gamma_q(S^q)$ corresponding to

$$S^q \xrightarrow{f} F\Gamma_q \hookrightarrow B\Gamma_q .$$

Since $B\nu \circ f$ is constant, γ has trivial normal bundle. γ can be represented by a foliation on a suitable open neighborhood of S^q in R^{q+1}. The submersion theorem [3] of Phillips then allows (cf. [1, p.192]) γ to be extended to a Γ_q-structure on D^{q+1}. That is, one finds $\omega \in \Gamma_q(D^{q+1})$ such that

$$\gamma = i^*(\omega)$$

where

$$i : S^q \to D^{q+1}$$

is the inclusion. ω has trivial normal bundle since every vector bundle over a contractible space is trivial. Thus $f : S^q \to F\Gamma_q$ extends to a map $\tilde{f} : D^{q+1} \to F\Gamma_q$, which proves $\pi_q(F\Gamma_q) = 0$.

Next we take up the exotic or "secondary" characteristic classes of foliations, the existence of which will be seen to be another consequence of the proof of (*). [Redactor's note. At the time these lectures were given the theory of these characteristic classes had not come into proper focus. Indeed, only the class $\omega(E)$ (cf.(10.7)) was discussed in the lectures and the existence of a foliation E with $\omega(E) \neq 0$ was left as an open question. Later in the summer a Comptes Rendus note by Godbillon-Vey appeared and furnished an example (due to Roussarie -- cf. (10.8)) of such an E. Much of what follows was developed in a seminar conducted by Milnor and Bott in the late summer of 1971 at La Jolla and later, during the fall of 1971, at the Institute for Advanced Study in Princeton.]

We begin with an example. Let $E \subset T(M)$ be a foliation of codimension one. Assume that the normal bundle Q is trivial. Then we can write $E = \mathrm{Ker}(\theta)$ where $\theta \in A^1(M)$ is nowhere zero. Furthermore, by (1.3) we can write

$$d\theta = \theta \cdot \eta$$

for suitable $\eta \in A^1(M)$.

The form θ is a trivialization of Q^*, hence under a choice of decomposition

$$\mathfrak{X} = \Gamma(E) \oplus \Gamma(Q)$$

there is a unique $Z \in \Gamma(Q)$ such that

$$\theta(Z) \equiv 1.$$

Relative to this global section of Q any connection ∇ on Q has a corresponding connection form φ such that

$$\nabla_X(Z) = \varphi(X)Z, \ \forall\, X \in \mathfrak{X}.$$

(10.4) Lemma. $\eta \in A^1(M)$ satisfies $d\theta = \theta \cdot \eta$ iff η is the connection form relative to Z of a basic connection on Q.

Proof. Let φ be the connection form relative to Z of a basic connection. Then $\varphi' \in A^1(M)$ is another such iff $\varphi|\Gamma(E) = \varphi'|\Gamma(E)$. Let $X \in \Gamma(E)$. Then, if $d\theta = \theta \cdot \eta$,

$$\begin{aligned} -\tfrac{1}{2}\,\eta(X) &= (\theta \cdot \eta)(X,Z) = d\theta(X,Z) \\ &= -\tfrac{1}{2}\,\theta([X,Z]) + \tfrac{1}{2}\,X(\theta(Z)) - \tfrac{1}{2}\,Z(\theta(X)) \\ &= -\tfrac{1}{2}\,\theta([X,Z]) = -\tfrac{1}{2}\,\theta(\varphi(X)Z) = -\tfrac{1}{2}\,\varphi(X), \end{aligned}$$

where we have used the standard formula [2,p.36] for the exterior derivative of a 1-form. Thus $\eta|\Gamma(E) = \varphi|\Gamma(E)$, so η is the form relative to Z of a basic connection.

For the converse, we have by the above that

$$d\theta(X,Z) = -\frac{1}{2}\varphi(X) = (\theta\cdot\varphi)(X,Z)$$

whenever $X \in \Gamma(E)$. If $X, Y \in \Gamma(E)$, then $[X,Y] \in \Gamma(E)$, so the formula for $d\theta$ shows $d\theta(X,Y) = 0$. Finally, $d\theta(Z,Z) = 0$, so

$$d\theta = \theta\cdot\varphi.$$

q.e.d.

(10.5) Lemma. Let η be as in (10.4). Then $\eta\cdot d\eta$ is a closed form. Furthermore, $[\eta\cdot d\eta] \in H^3_{DR}(M)$ is independent of the choice of η.

Proof. $d\eta = k$, the curvature form of a basic connection, by the structure equation (4.5). Thus

$$d(\eta\cdot d\eta) = d\eta\cdot d\eta = k^2 .$$

By the proof of (*), $k^2 = 0$.

If $d\theta = \theta\cdot\eta = \theta\cdot\eta'$, then $\eta' = \eta + f\theta$ for a suitable smooth function f. Thus

$$\eta'\cdot d\eta' = \eta\cdot d\eta + \eta\cdot d(f\theta) + f\theta\cdot d\eta + f\theta\cdot d(f\theta).$$

But

$$f\theta\cdot d(f\theta) = f\theta\cdot df\cdot\theta + f^2\theta\cdot d\theta$$
$$= f^2\theta\cdot d\theta = f^2\theta\cdot\theta\cdot\eta = 0$$

and

$$\theta\cdot d\eta = \theta\cdot d\eta - \theta\cdot\eta^2 = d(-\theta\cdot\eta) = -d^2(\theta) = 0 ,$$

so,

$$\eta'\cdot d\eta' = \eta\cdot d\eta + \eta\cdot d(f\theta) - f\theta\cdot d\eta = \eta\cdot d\eta - d(\eta\cdot f\theta) .$$

q.e.d.

(10.6) Exercise. Show that $[\eta \cdot d\eta]$ is also independent of the choice of trivialization θ of Q^*.

(10.7) Definition. If $E \subset T(M)$ is a foliation of codimension one having trivial normal bundle, then $\omega(E) = [\eta \cdot d\eta] \in H^3_{DR}(M)$.

Remark that the exotic characteristic class $\omega(E)$ is defined because of the vanishing of k^2. By the above, it depends only on the foliation E (indeed, by standard homotopy arguments, only on the class of E in $\mathrm{Fol}_1(M)$). The fact that it need not always be zero is established by an example due to Roussarie which we now describe.

Let $SL(2,R)$ denote the subgroup of GL_2 consisting of the elements of determinant 1. It is well known that this group admits discrete subgroups G such that the right coset space

$$M = G\backslash SL(2,R)$$

is a compact manifold. Since $SL(2,R)$ has dimension three, so does M.

(10.8) Theorem. (Roussarie) M as above admits a codimension one foliation $E \subset T(M)$ such that $\omega(E) \neq 0$.

Proof. Let $K \subset SL(2,R)$ be the two dimensional subgroup of matrices

$$\begin{pmatrix} a & 0 \\ b & c \end{pmatrix}, \quad ac = 1 .$$

Thus $SL(2,R)$ is foliated by the left cosets of K, a foliation invariant under the action of G from the left with trivial normal bundle also invariant under G. The Lie algebra of left invariant vector fields on $SL(2,R)$ identifies naturally with the Lie algebra of real matrices

$$\begin{pmatrix} x & y \\ z & w \end{pmatrix}, \quad x + w = 0 .$$

The Lie algebra of left invariant vector fields tangent to the above foliation corresponds to the subalgebra of the above set of matrices in which $y = 0$. Thus $\Gamma(E)$ is spanned by two left invariant vector fields X and Y corresponding to matrices as follows

$$X \leftrightarrow \begin{pmatrix} 1 & 0 \\ 0 & -1 \end{pmatrix}, \quad Y \leftrightarrow \begin{pmatrix} 0 & 0 \\ 1 & 0 \end{pmatrix}$$

while there is a trivializing section $Z \in \Gamma(Q)$ such that

$$Z \leftrightarrow \begin{pmatrix} 0 & 1 \\ 0 & 0 \end{pmatrix} .$$

The brackets of these fields are found by taking the commutators of the corresponding matrices. Thus

$$[X,Y] = -2Y$$
$$[X,Z] = 2Z$$
$$[Y,Z] = -X .$$

Let X^*, Y^*, Z^* be the dual basis of 1-forms. Then Z^* is the θ of our previous discussion and, by the formula for $d\theta$,

$$d\theta(X,Y) = -\frac{1}{2} Z^*([X,Y]) = 0$$

$$d\theta(X,Z) = -\frac{1}{2} Z^*([X,Z]) = -1$$

$$d\theta(Y,Z) = -\frac{1}{2} Z^*([Y,Z]) = 0$$

since the terms $\frac{1}{2} X(Z^*(Y))$, $\frac{1}{2} Y(Z^*(X))$, etc. all vanish. Thus

$$d\theta = -X^* \cdot Z^* = \theta \cdot X^* ,$$

so the η of (10.4) is X^*. Computations exactly like the above show

$$d\eta = \frac{1}{2} Y^* \cdot Z^*$$

and so

$$\eta \cdot d\eta = \frac{1}{2} X^* \cdot Y^* \cdot Z^* .$$

This is a volume element (i.e., a nowhere zero 3-form) on $SL(2,R)$ since X^*, Y^*, Z^* are everywhere linearly independent. All of these constructions and definitions are invariant under the left action of G, hence carry over to M. On M the $\eta \cdot d\eta$ corresponding to the foliation is also a volume element. Since M is compact, it is a well known fact that $[\eta \cdot d\eta] \in H^3_{DR}(M)$ is nonzero. q.e.d.

As a first step toward generalizing the class $\omega(E)$, we show how it can be obtained by comparing two connections on Q. Indeed, quite generally let Q be any smooth q-dimensional vector bundle over an m-dimensional manifold M and let $\nabla^0, \nabla^1, \ldots, \nabla^n$ be connections on Q. Let $\nabla^{0,1,\ldots,n}$ be the connection on the bundle

$$Q \times R^n \to M \times R^n$$

obtained by affine combinations

$$(1-a_1 - \cdots - a_n)\, \nabla^0 + a_1\, \nabla^1 + \cdots + a_n \nabla^n$$

as in § 5. Let $I(gl_q)$ denote the ring of invariant polynomials on gl_q (cf. (5.1)) and define

$$\lambda(\nabla^0, \nabla^1, \ldots, \nabla^n) : I(gl_q) \to A^*_C(M)$$

as follows. For the case $n = 0$, let

$$\lambda(\nabla^0)(\varphi) = \left(\frac{\sqrt{-1}}{2\pi}\right)^{\deg(\varphi)} \varphi(k^0)$$

for each $\varphi \in I(gl_q)$, where k^0 is the curvature of ∇^0. For the general case, set

$$\lambda(\nabla^0, \ldots, \nabla^n)(\varphi) = (-1)^{[n/2]}\, \pi_*(\lambda(\nabla^{0,\ldots,n})(\varphi) \mid M \times \Delta^n)$$

where

$$\pi_* : A^*(M \times \Delta^n) \to A^*(M)$$

is integration along the fiber (cf. (3.10)).

Note that for $n = 0$ this is a ring homomorphism. Also note that $\lambda(\nabla^0,\ldots,\nabla^n)(\varphi)$ is real for $\deg(\varphi)$ even and pure imaginary for $\deg(\varphi)$ odd. Finally, remark that $\lambda(\nabla^0,\ldots,\nabla^n)$ is antisymmetric in the $n+1$ entries.

As in § 5, we use the combinatorial version of Stoke's Theorem and the relation (3.10) together with the fact that $\lambda(\nabla^{0,\ldots,n})(\varphi)$ is a closed form (cf. (5.3)) to verify the formula

$$(\#) \qquad d(\lambda(\nabla^0,\ldots,\nabla^n)(\varphi)) = \sum_{i=0}^{n} (-1)^i \lambda(\nabla^0,\ldots,\hat{\nabla}^i,\ldots,\nabla^n)(\varphi) .$$

Indeed, the affine combination of the connections ∇^i was so set up that

$$\pi_*^{\partial}(\lambda(\nabla^{0,\ldots,n})(\varphi)\,|\,M \times \partial\Delta^n) = \sum_{i=0}^{n} (-1)^i \; \pi_*(\lambda(\nabla^{0,\ldots,\hat{i},\ldots,n})(\varphi)\,|\,M \times \Delta^{n-1}) ,$$

hence the formula(3.10)

$$(\pi_* \circ d + (-1)^{n+1} d \circ \pi_*)(\lambda(\nabla^{0,\ldots,n})(\varphi)\,|\,M \times \Delta^n) = \pi_*^{\partial}(\lambda(\nabla^{0,\ldots,n})(\varphi)\,|\,M \times \partial\Delta^n)$$

gives

$$d(\lambda(\nabla^0,\ldots,\nabla^n)(\varphi)) = (-1)^{[n/2]} d\{\pi_*(\lambda(\nabla^{0,\ldots,n})(\varphi)\,|\,M \times \Delta^n)\}$$

$$= (-1)^{n+1+[n/2]} \; \pi_*^{\partial}(\lambda(\nabla^{0,\ldots,n})(\varphi)\,|\,M \times \partial\Delta^n)$$

$$= (-1)^{n+1+[n/2]+[(n-1)/2]} \sum_{i=0}^{n} (-1)^i \lambda(\nabla^0,\ldots,\hat{\nabla}^i,\ldots,\nabla^n)(\varphi) .$$

But

$$n + 1 + [n/2] + [(n-1)/2] = 2n$$

so we obtain (#).

Returning to $\omega(E)$, let $E \subset T(M)$ be a foliation of codimension one with trivial normal bundle Q. Let $Z \in \Gamma(Q) \subset \mathcal{X}$ be a trivializing section, θ the corresponding 1-form, $d\theta = \theta \cdot \eta$, and let ∇^1 be the basic connection on Q for which

$$\nabla^1_X(Z) = \eta(X)Z, \ \forall X \in \mathfrak{X} .$$

Let ∇^0 be the connection on Q for which

$$\nabla^0_X(Z) = 0 , \ \forall X \in \mathfrak{X} .$$

Finally, note that $I(g\ell_1) = R[x]$.

(10.9) Lemma. The following three formulas hold.

1) $\lambda(\nabla^1)(x) = (\frac{\sqrt{-1}}{2\pi})d\eta$

2) $\lambda(\nabla^0,\nabla^1)(x) = (\frac{\sqrt{-1}}{2\pi})\eta$

3) $\lambda(\nabla^0,\nabla^1)(x^2) = (\frac{-1}{4\pi^2})\eta \cdot d\eta$.

Proof. Since $d\eta$ is the curvature of ∇^1, the first formula is immediate. If $\pi : M \times R \longrightarrow M$ is the standard projection, then a trivial computation shows that the connection form of $\nabla^{0,1}$ is

$$\psi = t\,\pi^*(\eta) .$$

Thus the curvature form is

$$k^{0,1} = d\psi = dt \cdot \pi^*(\eta) + t \cdot \pi^*(d\eta).$$

The square of this 2-form is

$$2t\, dt \cdot \pi^*(\eta \cdot d\eta) + t^2\, \pi^*(d\eta)^2 .$$

Plugging these into our formula gives 2) and 3). q.e.d.

(10.10) Corollary. The class $\omega(E)$ is $-4\pi^2$ times the cohomology class of the closed 3-form

$$\lambda(\nabla^0,\nabla^1)(x^2) = \lambda(\nabla^0,\nabla^1)(x) \cdot \lambda(\nabla^1)(x) .$$

Noting that ∇^0 is a Riemannian connection (it preserves the metric in which $\|Z\| = 1$) we use (10.10) to motivate a general definition of secondary characteristic classes for foliations. These will all be obtained by comparing a Riemannian connection and a basic connection on the normal bundle Q and will be independent of the choices of connections.

(10.11) Theorem. $I(g\ell_q) = R[c_1, c_2, \ldots, c_q]$ where the invariant polynomials c_i are defined by

$$\det(I + tA) = 1 + \sum_{i=1}^{q} t^i c_i(A) \quad .$$

For a proof, cf. the appendix.

If $E \subset T(M)$ is a codimension q foliation, $Q = T(M)/E$, and if ∇^1 is a basic connection on Q, then by (*)

$$\lambda(\nabla^1) : R[c_1, \ldots, c_q] \to A_C^*(M)$$

annihilates all elements of degree $> q$.

Thus, if

$$R_q[c_1, \ldots, c_q] = R[c_1, \ldots, c_q]/\{\varphi : \deg(\varphi) > q\}$$

it follows that $\lambda(\nabla^1)$ defines

$$\lambda(\nabla^1) : R_q[c_1, \ldots, c_q] \to A_C^*(M) \; .$$

On the other hand, if ∇^0 is a Riemannian connection on Q, then the argument for (5.6) shows that

$$\lambda(\nabla^0) : R[c_1, \ldots, c_q] \to A_C^*(M)$$

annihilates elements of odd degree. In particular,

$$\lambda(\nabla^0)(c_{2i-1}) = 0, \quad 1 \le 2i-1 \le q \quad .$$

Thus, application of (#) shows

$$\text{(10.12)} \qquad d\{\lambda(\nabla^0,\nabla^1)(c_{2i-1})\} = \lambda(\nabla^1)(c_{2i-1}).$$

These relations enable us to build a cochain complex WO_q and a homomorphism of cochain complexes

$$\lambda_E : WO_q \to A_{\mathbb{C}}^*(M) .$$

As a graded algebra,

$$WO_q = R_q[c_1,\ldots,c_q] \otimes E(h_1,h_3,\ldots,h_\ell)$$

where ℓ is the largest odd integer $\leq q$, each c_i is assigned degree $2i$, and each h_i is assigned degree $2i-1$. Here, of course, $E(h_1,\ldots,h_\ell)$ designates the exterior algebra over R generated by the elements h_i. A unique antiderivation of degree 1

$$d : WO_q \to WO_q$$

is defined by requiring

$$d(c_i) = 0, \quad 1 \leq i \leq q$$

$$d(h_i) = c_i, \quad i = 1,3,\ldots,\ell \quad .$$

Clearly $d^2 = 0$. The unique homomorphism of graded R-algebras

$$\lambda_E : WO_q \to A_{\mathbb{C}}^*(M)$$

is defined by requiring

$$\lambda_E(c_i) = \lambda(\nabla^1)(c_i), \quad 1 \leq i \leq q ,$$

$$\lambda_E(h_i) = \lambda(\nabla^0,\nabla^1)(c_i), \quad i = 1,3,\ldots,\ell .$$

By (10.12) and the fact that $\lambda(\nabla^1)(c_i)$ is a closed form,

$$\lambda_E \circ d = d \circ \lambda_E \quad .$$

Thus, in cohomology λ_E induces a homomorphism of graded $\mathbb{R}$-algebras

$$\lambda_E^* : H^*(WO_q) \to H_{DR}^*(M;C).$$

λ_E depends, of course, on the choices of ∇^0 and ∇^1, but λ_E^* does not.

(10.13) Proposition. λ_E^* depends only on the foliation E, not on the choices of Riemannian connection ∇^0 and basic connection ∇^1.

Proof. If ∇^1 and $\bar{\nabla}^1$ are basic connections, then $(1-t)\nabla^1 + t\bar{\nabla}^1$ is also a basic connection, $0 \leq t \leq 1$. Similarly the set of Riemannian metrics on Q is a convex set, so any two Riemannian connections are readily seen to be homotopic through Riemannian connections. Thus, as in the proof of (5.4), the assertion follows by the homotopy invariance of $H_{DR}^*(M;C)$. q.e.d.

A notion of homotopy between codimension q foliations E and E' on M which is weaker than (9.1) is obtained simply by requiring a codimension q foliation F on $M \times \mathbb{R}$ such that $M \times \{0\}$ and $M \times \{1\}$ are transverse to F and $i_0^*(F) = E$, $i_1^*(F) = E'$. Then, for purely homotopy theoretic reasons, we obtain

(10.14) Proposition. λ_E^* depends only on the homotopy class of E.

The homomorphism $\lambda_E^* : H^*(WO_q) \to H_{DR}^*(M;C)$ has a certain naturality property. Indeed, if $N \xrightarrow{i} M$ is a submanifold transversal to the leaves of the foliation E, there is induced a codimension q foliation $E' = i^*(E)$ on N. The normal bundle of E' is just $Q|N = i^*(Q)$ and both Riemannian and basic connections on Q restrict to Riemannian and basic connections respectively on $i^*(Q)$.

(10.15) Proposition. Under the above hypotheses, the diagram

$$\begin{array}{ccc} H^*(WO_q) & \xrightarrow{\lambda_E^*} & H^*_{DR}(M;C) \\ & \searrow^{\lambda^*_{E'}} & \downarrow i^* \\ & & H^*_{DR}(N;C) \end{array}$$

is commutative.

As a consequence of this naturality we can define a canonical homomorphism

$$\lambda^* : H^*(WO_q) \to H^*(B\Gamma_q;C)$$

where we use singular cohomology on $B\Gamma_q$. Indeed, if $\alpha \in H^r(WO_q)$, we can determine

$$\lambda^*(\alpha) \in H^r(B\Gamma_q;C) = \mathrm{Hom}_C(H_r(B\Gamma_q;C),C)$$

by showing how to evaluate it on homology classes $[\sigma] \in H_r(B\Gamma_q;C)$. As in the proof of (10.1), there is an open manifold M and a codimension q foliation E of M, the classifying map of which

$$s : M \to B\Gamma_q$$

satisfies

$$[\sigma] = s_*[\sigma']$$

for suitable $[\sigma'] \in H_r(M;C)$. Indeed, M is a neighborhood of a polyhedron P in a suitable Euclidean space such that P is a deformation retract of M and

$$s : P \to B\Gamma_q$$

is the geometric realization of the cycle σ. Then define

$$\lambda^*(\alpha)[\sigma] = \lambda_E^*(\alpha)[\sigma'] .$$

Here, of course, we assume the de Rham theorem for singular cohomology [4].

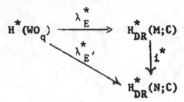

In order to see that $\lambda^*(\alpha)[\sigma]$ is well defined, let R be another polyhedron with map

$$u : R \longrightarrow B\Gamma_q$$

which is the geometric realization of a cycle $\bar{\sigma} = u_\#(\sigma'')$ homologous to σ. Let $R \subset N$ be a deformation retract, N an open manifold with codimension q foliation F classified by the extension of u to N. One then builds a polyhedron K and a map

$$v : K \longrightarrow B\Gamma_q$$

such that P and R are subpolyhedra and $s = v|P$, $u = v|K$, and such that the cycles σ' and σ'' are homologous in K. Let $K \subset W$ as usual, W an open manifold with codimension q foliation G classified by v. Without loss of generality, we suppose M and N open submanifolds of W, the foliations E and F respective restrictions of G, the maps s and u respective restrictions of v. Thus the naturality (10.15) gives a commutative diagram

$$\begin{array}{ccc}
 & \overset{\lambda_E^*}{\nearrow} & H^r_{DR}(M;C) \\
 & & \uparrow i^* \\
H^r(WO_q) & \xrightarrow{\lambda_G^*} & H^r_{DR}(W;C) \\
 & & \downarrow j^* \\
 & \underset{\lambda_F^*}{\searrow} & H^r_{DR}(N;C)
\end{array}$$

where i and j are the inclusions. Since $i_*[\sigma'] = j_*[\sigma'']$ in $H_r(W;C)$, we have

$$\begin{aligned}
\lambda_E^*(\alpha)[\sigma'] &= i^*\lambda_G^*(\alpha)[\sigma'] \\
&= \lambda_G^*(\alpha)\{i_*[\sigma']\} = \lambda_G^*(\alpha)\{j_*[\sigma'']\} \\
&= j^*\lambda_G^*(\alpha)[\sigma''] = \lambda_F^*(\alpha)[\sigma''] .
\end{aligned}$$

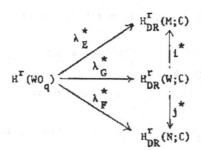

(10.16) **Theorem.** There is a canonical homomorphism

$$\lambda^* : H^*(WO_q) \longrightarrow H^*(B\Gamma_q;C)$$

with the property that, for any foliation $E \subset T(M)$ with classifying map $g : M \longrightarrow B\Gamma_q$, the diagram

$$\begin{array}{ccc} H^*(WO_q) & \xrightarrow{\lambda^*} & H^*(B\Gamma_q;C) \\ & \searrow^{\lambda_E^*} & \downarrow{g^*} \\ & & H^*(M;C) \end{array}$$

is commutative.

$Im(\lambda^*)$ is, then, the desired set of universal exotic characteristic classes for foliations (actually, for Γ_q-structures).

The extent to which $\lambda^* \neq 0$ is of fundamental interest, and for this Roussarie's example plays a decisive role. In WO_1, the elements $1, c_1, h_1$, and c_1h_1 are an R-linear basis, and the only cycles are the linear combinations of 1, c_1, and c_1h_1. $c_1 = d(h_1)$, hence

$$H^r(WO_1) = \begin{cases} R, & r = 0,3 \\ 0, & \text{otherwise} \end{cases}$$

and the generator in $H^3(WO_1)$ is $[c_1h_1]$. By (10.9), if $E \subset T(M)$ is a codimension one foliation with trivial Q, then

$$\omega(E) = (-4\pi^2)\lambda_E^*[c_1h_1] .$$

By (10.8) we obtain

(10.17) **Theorem.** (Roussarie) $\lambda^*[c_1h_1] \neq 0$ in $H^3(B\Gamma_1;C)$.

Remark that, in defining $\lambda^*[c_1h_1]$ we used not only the codimension one hypothesis, but restricted ourselves to foliations with trivial normal bundle. The classifying map

$$g : M \longrightarrow B\Gamma_1$$

for such a foliation factors through $F\Gamma_1$, hence the above technique provides a nontrivial element in $H^3(F\Gamma_1;C)$. With this clue we can modify the definition of λ^* to obtain a canonical homomorphism

$$\lambda^* : H^*(W_q) \longrightarrow H^*(F\Gamma_q;C)$$

for a suitable cochain complex W_q. Indeed, considering only codimension q foliations $E \subset T(M)$ with trivial normal bundle, we replace the general Riemannian connection ∇^0 in the above construction with a <u>flat</u> connection ∇^0 (i.e., one for which Q admits independent global sections $s_1,\ldots,s_q$ with $\nabla^0_X(s_i) = 0$, $\forall X \in \mathfrak{X}$). Then $\lambda(\nabla^0)(c_i) = 0$ for all i, so we define

$$W_q = R_q[c_1,\ldots,c_q] \otimes E(h_1,h_2,\ldots,h_q)$$

with $\deg(c_i) = 2i$, $\deg(h_i) = 2i-1$, and $d(c_i) = 0$, $d(h_i) = c_i$. The homomorphism

$$\lambda_E : W_q \rightarrow A^*_C(M)$$

is defined by

$$\lambda_E(c_i) = \lambda(\nabla^1)(c_i)$$

$$\lambda_E(h_i) = \lambda(\nabla^0,\nabla^1)(c_i)$$

for ∇^0 a flat connection, ∇^1 a basic connection. One obtains the desired homomorphism of $H^*(W_q)$ into $H^*(F\Gamma_q;C)$ by mimicry of the previous techniques.

In conclusion, we mention an application of the analogue of $\omega(E)$ in the theory of holomorphic foliations. Here we study a fibration

$$F\Gamma_q C \longrightarrow B\Gamma_q C \xrightarrow{B\nu} BGL_q C$$

where $\Gamma_q C$ is the topological category with C^q as the space of objects and with morphisms the sheaf of germs of local holomorphic diffeomorphisms of C^q. The holomorphic analogue of $\omega(E)$ gives a universal class

$$\omega \in H^3(F\Gamma_1 C;C) .$$

Let $M = C^2 - \{0\}$, an open complex analytic manifold having the homotopy type of S^3. Then the set of homotopy classes of maps $[M,F\Gamma_1 C]$ canonically identifies with $\pi_3(F\Gamma_1 C)$. Let $\sigma \in H_3(M;C)$ be the homology class represented by the unit sphere $S^3 \subset M$. The class ω then defines a map

$$\omega_* : \pi_3(F\Gamma_1 C) \longrightarrow C$$

by

$$\omega_*([f]) = \{f^*(\omega)\}(\sigma)$$

for each $[f] \in [M,F\Gamma_1 C]$.

On M consider the nowhere zero holomorphic 1-form

$$\theta_{\alpha\beta} = \alpha z_2 dz_1 + \beta z_1 dz_2$$

where α and β are nonzero complex constants. $\mathrm{Ker}(\theta_{\alpha\beta})$ is a one dimensional holomorphic subbundle of the holomorphic tangent bundle $T(M)$, hence is integrable, and the foliation has complex codimension one with trivial complex normal bundle. From the identity

$$d\theta_{\alpha\beta} = (\beta-\alpha)\, dz_1 \cdot dz_2$$

$$= \theta_{\alpha\beta} \cdot \left\{ \left(\frac{\beta-\alpha}{|\alpha z_2|^2 + |\beta z_1|^2} \right) (\overline{\alpha}\,\overline{z}_2 dz_2 - \overline{\beta}\,\overline{z}_1 dz_1) \right\}$$

$$= \theta_{\alpha\beta} \cdot \eta_{\alpha\beta}$$

we obtain the connection form $\eta_{\alpha\beta}$ for a suitable complex C^∞ basic connection. Since $\eta_{\alpha\beta}$ is of the form $f\phi$ where

$$f = \frac{\beta-\alpha}{|\alpha z_2|^2 + |\beta z_1|^2}$$

$$\phi = \overline{\alpha z}_2 dz_2 - \overline{\beta z}_1 dz_1 \quad ,$$

we obtain

$$\eta_{\alpha\beta} \cdot d\eta_{\alpha\beta} \qquad f^2\phi \cdot d\phi = \overline{\alpha\beta}\, f^2 \psi \cdot dz_1 \cdot dz_2$$

where

$$\psi = \overline{z}_1 d\overline{z}_2 - \overline{z}_2 d\overline{z}_1 \quad .$$

the change of variables

$$u = \alpha z_2$$

$$v = \beta z_1$$

we obtain

$$\eta_{\alpha\beta} \cdot d\eta_{\alpha\beta} = \left(\frac{\beta}{\alpha} + \frac{\alpha}{\beta} - 2 \right) \left\{ \frac{(\overline{v}\, d\overline{u} - \overline{u}\, d\overline{v}) \cdot dv \cdot du}{(|u|^2 + |v|^2)^2} \right\} \quad .$$

The unit sphere relative to the (u,v)-coordinates is homotopic to the unit sphere relative to the (z_1, z_2)-coordinates, hence evaluation of $[\eta_{\alpha\beta} \cdot d\eta_{\alpha\beta}]$ on the corresponding hemology class becomes

$$\int_S \eta_{\alpha\beta} \cdot d\eta_{\alpha\beta} = \left(\frac{\beta}{\alpha} + \frac{\alpha}{\beta} - 2 \right) \int_{S^3} (\overline{v}\, d\overline{u} - \overline{u}\, d\overline{v}) \cdot dv \cdot du \quad .$$

Letting

$$u = x_1 + ix_2$$

$$v = x_3 + ix_4$$

one shows that on S^3 the integrand on the right is a real nowhere zero 3-form, hence

$$\int_{S^3} \eta_{\alpha\beta} \cdot d\eta_{\alpha\beta} = c\left(\frac{\beta}{\alpha} + \frac{\alpha}{\beta} - 2 \right)$$

where c is a nonzero real constant independent of α and β. By suitable choices of α and β this expression can be made to assume any complex value γ. Indeed, set $\alpha = 1$ and solve the resulting equation

$$c\beta^2 - (\gamma + 2c)\beta + c = 0$$

for nonzero β. If

$$f_{\alpha\beta} : M \to F\Gamma_1 C$$

corresponds to the foliation in question, then $\omega_*([f_{\alpha\beta}])$ assumes any complex value γ by suitable choices of α and β.

(10.18) Theorem. $\omega_* : \pi_3(F\Gamma_1 C) \to C$ is a surjection.

References

1. A. Haefliger, Feuilletages sur les variétés ouvertes, Topology 9 (1970), pp. 183-194.

2. S. Kobayashi and K. Nomizu, Foundations of Differential Geometry, Vol. I, 1963, Interscience Publishers, New York.

3. A. Phillips, Submersions of open manifolds, Topology 6 (1967), 171-206.

4. A. Weil, Sur les théorèmes de de Rham, Comment. Math. Helv., 26 (1952), pp. 119-145.

__Appendix__ A. We discuss the algebra of invariant polynomials. Denote by F the field $\mathbb{R}$ or $\mathbb{C}$ and by $I_q(F)$ the algebra of polynomials on $\mathfrak{gl}(q,F)$ invariant under conjugation by $GL(q,F)$. Let $S_q(F)$ denote the algebra of symmetric polynomials in q variables over F, and let $\sigma_i \in S_q(F)$ be the ith elementary symmetric function. It is well known that

$$S_q(F) = F[\sigma_1,\sigma_2,\ldots,\sigma_q],$$

a polynomial algebra (cf. [2,p. 177]).

Given $\varphi \in I_q(F)$, the restriction of φ to the set of diagonal matrices produces a symmetric polynomial in the diagonal entries. Indeed, any permutation of these diagonal entries can be produced via conjugation by suitable elements of $GL(n,F)$. This defines a canonical homomorphism of algebras

$$\rho : I_q(F) \to S_q(F) .$$

Lemma A. ρ is surjective.

Proof. Recall the elements $c_i \in I_q(F)$ defined by the formula

$$\det(I + tA) = 1 + \sum_{i=1}^{q} t^i c_i(A) .$$

The restriction of c_i to the set of diagonal matrices is simply the ith elementary symmetric function. That is, $\rho(c_i) = \sigma_i$, $1 \le i \le q$, so ρ is surjective. q.e.d.

Lemma B. $\rho : I_q(\mathbb{C}) \to S_q(\mathbb{C})$ is injective.

Proof. Let $\Delta_q(\mathbb{C})$ denote the vector space of upper triangular $q \times q$ matrices over $\mathbb{C}$. Since a matrix is diagonalizable iff its minimal polynomial is a product of distinct linear factors [1,p. 175] it follows that the subset $\Delta_q^*(C) \subset \Delta_q(C)$ consisting of those matrices whose diagonal entries are pairwise distinct is a set of diagonalizable matrices. Clearly $\Delta_q^*(C)$ is dense in $\Delta_q(C)$, hence the subset of all diagonalizable matrices in $\Delta_q(C)$ is dense.

It is an elementary fact that every matrix in $gl(q,C)$ is conjugate under $GL(q,C)$ to an upper triangular matrix. Indeed, if A is a linear transformation of C^q, the algebraic closure of C guarantees the existence of a nonzero eigenvector $v_1 \in C^q$ for A. A defines a linear transformation of $V_1 = C^q/\mathrm{span}(v_1)$, hence has a nonzero eigenvector $\bar{v}_2 \in V_1$. This gives $v_2 \in C^q$ linearly independent of v_1 such that $A(v_2) \in \mathrm{span}(v_1,v_2)$. Proceeding inductively, produce a basis $v_1,v_2,\ldots,v_q$ of C^q such that $A(v_i) \in \mathrm{span}(v_1,\ldots,v_i)$, $1 \leq i \leq q$. Relative to this basis A is represented by an upper triangular matrix.

By these remarks, the set of all $q \times q$ diagonalizable matrices is dense in $gl(q,C)$. If $\varphi \in I_q(C)$ and $\rho(\varphi) = 0$, it follows that φ vanishes on every diagonalizable matrix. By continuity and denseness, $\varphi = 0$. q.e.d.

Lemma C. $\rho : I_q(R) \longrightarrow S_q(R)$ is injective.

Proof. For purely formal reasons, every $\varphi \in I_q(R)$ is also invariant when interpreted as a polynomial on $gl(q,C)$. This gives an inclusion of $I_q(R)$ as a real subalgebra of $I_q(C)$. Likewise, $S_q(R)$ is a real subalgebra of $S_q(C)$. The diagram

$$\begin{array}{ccc} I_q(R) & \subset & I_q(C) \\ \rho \downarrow & & \downarrow \rho \\ S_q(R) & \subset & S_q(C) \end{array}$$

is commutative, so Lemma B gives the assertion. q.e.d.

These lemmas establish (10.11).

Finally, recall the elements $\Sigma_i \in I_q(F)$ defined by

$$\Sigma_i(A) = \text{trace}(A^i) .$$

Then

$$\rho(\Sigma_i)(x_1,\ldots,x_q) = x_1^i + x_2^i + \cdots + x_q^i .$$

Lemma D. The elements $\{\Sigma_1,\Sigma_2,\ldots,\Sigma_r,\ldots\}$ generate $I_q(F)$.

Proof. Since ρ is an isomorphism, it is enough to prove that the symmetric polynomials $\{\rho(\Sigma_1),\rho(\Sigma_2),\ldots,\rho(\Sigma_r),\ldots\}$ generate $S_q(F)$. This will follow by induction on k from the classical formula (in which we use the convention $\sigma_0 = 1$)

$$0 = k\sigma_k + \sum_{i=1}^{k} (-1)^i \sigma_{k-i}\, \rho(\Sigma_i),\ 1 \leq k \leq q .$$

In order to prove this formula, first define

$$f = 1 + \sigma_1 t + \cdots + \sigma_q t^q \in S_q(F)[t] .$$

That is,

$$f = (1 + x_1 t)(1 + x_2 t)\cdots(1 + x_q t).$$

Then, for purely formal reasons,

$$- t\left(\frac{df}{dt}\right)/f = - t \frac{d}{dt}(\log f(t))$$

$$= - tx_1/(1 + x_1 t) - \cdots - tx_q/(1 + x_q t)$$

$$= \sum_{i \geq 1} (x_1^i + \cdots + x_q^i)(-t)^i = \sum_{i \geq 1} (-1)^i \rho(\Sigma_i) t^i .$$

Multiplying both sides by f gives

$$\left(\sum_{j \geq 0} \sigma_j t^j\right)\left(\sum_{i \geq 1} (-1)^i \rho(\Sigma_i) t^i\right) = - t \sum_{k \geq 0} k\sigma_k t^{k-1}$$

and so

$$\sum_{k \geq 1} \{ (\sum_{i+j=k} (-1)^i \sigma_j \rho(\Sigma_i)) + k\sigma_k \} t^k = 0 .$$

Here, of course, we also use the convention $\sigma_j = 0$ if $j > q$. q.e.d.

References

1. K. Hoffman and R. Kunze, Linear Algebra, Prentice-Hall, Englewood Cliffs, N. J., 1961.

2. D. Husemoller, Fibre Bundles, McGraw-Hill, New York, 1966.

Appendix B - Construction of BC

by J. Stasheff

The history of classifying spaces for groups is extensive [9]. Here we will study several important examples and compare alternative forms of the construction.

If G is a topological group, we can form the associated category $\mathcal{G}$ with one object $*$ and $\mathrm{Mor}(*,*) = G$, composition being given by $g \circ g' = g'g$, to conform with established usage. Segal's $B\mathcal{G}$ is then exactly BG as defined by Milgram [5] or, in an obscure exposition, by Stasheff [10, p. 289]. If $G = S^0$, S^1 or S^3 with the usual multiplication then BG is homeomorphic (respectively) to RP^∞, CP^∞ or HP^∞. If S^0, S^1 or S^3 is given the minimal cell structure, the induced cell structure on the projective space is the standard one. The identifications corresponding to the degeneracies s_i serve to achieve this minimality of the cell decomposition, but for our purposes they get in the way. The construction can be reworked as follows:

Recall that $A_0C = \mathrm{Ob}C$ and $A_nC \subset \mathrm{Mor}\, C \times \dots \times \mathrm{Mor}\, C$ is the subset consisting of n-tuples $(f_1,\dots,f_n)$ such that $f_{i+1} \circ f_i$ is defined. Let $\sigma = \sigma^n$ denote an $(n+1)$-tuple of integers $0 \leq i_0 \leq i_1 \leq \dots \leq i_n$, and Δ_σ the simplex with vertices in σ. Define

$$\mathcal{B}C = \coprod_{n,\sigma^n} \Delta_\sigma \times A_n/\sim$$

where the equivalence relation $\sim$ is defined by the face relations:

if $t_{i_k} = 0$,

$$(t_{i_0},\dots,t_{i_n}, f_1,\dots,f_n) \sim (t_{i_0},\dots,\hat{t}_{i_k},\dots,f_{k+1} \circ f_k,\dots).$$

This $\mathcal{B}$ is, like Segal's B, a functor from topological categories to topological spaces.

There is a map $\mathcal{B}C \to BC$ (as defined by Segal or on p. 61) which collapses the degenerate simplices, where any $A_n C$ coordinate is the identity. This is the standard reduction of a simplicial object to the normalized form, as it appears in singular theory or the Eilenberg-Zilber theorem [3, p. 236]. Just as there, it is a homotopy equivalence in this case [10, p. 289].

If G is a topological group and $\mathcal{G}$ is the associated category, we have $A_n\mathcal{G} = G^n$. The space $\mathcal{B}\mathcal{G}$ is identical with "BG" as defined by Dold and Lashof [2]. The point of their construction was that it did not use the inverses in G and so went through for a topological monoid (=associative H-space). For example, if ΩX is the (Moore) space of loops on a connected CW-complex X, then $\mathcal{B}\Omega X$ has the homotopy type of X. Closer to our interest, if $H(F)$ is the monoid of auto-homotopy - equivalences of F to itself, $\mathcal{B}H(F)$ classifies Hurewicz fibrations with fibre F over CW-base spaces [1,7].

For a group G, we also have Milnor's construction [6] which we will denote $\mathbb{B}G$. For comparison with $\mathcal{B}\mathcal{G}$, we note that $\mathbb{B}G$ can be defined by first taking

$$EG = \coprod_{n,\sigma^n} \Delta_\sigma \times G \times G^n/\approx$$

where the equivalence relation is: if $t_{i_k} = 0$,

$$(t_{i_0},\dots,t_{i_n},g_0,g_1,\dots,g_n) \approx (\dots,\hat{t}_{i_k},\dots,\hat{g}_k,\dots) .$$

Then define $\mathbb{B}G$ as the quotient by the action $g(g_0,\dots,g_n) = (gg_0,\dots,gg_n)$. (One can profitably think of homogeneous coordinates in projective geometry.).

The comparison between $\mathcal{B}\mathcal{G}$ and $\mathbb{B}G$ can be made quite explicit in terms of the isomorphism $\phi : G^n \to G^n$ given by

$$(g_1, g_2, \ldots, g_n) \to (g_1, g_1 g_2, \ldots, g_1 \cdots g_n) .$$

We then map $\mathcal{B}\mathcal{G} \to \mathbb{B}G$ by

$$(t_{i_0}, \ldots, t_{i_n}, g_1, \ldots, g_n) \mapsto (t_{i_0}, \ldots, t_{i_n}, e, \phi(g_1, \ldots, g_n)).$$

The map respects the identifications as we indicate on $\Delta^2 \times G^2$:

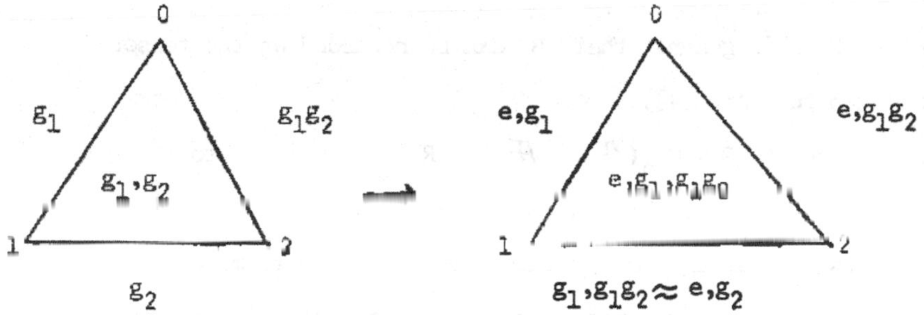

This isomorphism is the one which relates the standard inhomogeneous and homogeneous resolutions of a (abstract) group. The inverse to ϕ depends on the existence of inverses in G.

The variety of constructions available can be summarized as follows:

	homogeneous	inhomogeneous
normalized	unpublished	Milgram Stasheff Segal
unnormalized	Milnor	Dold-Lashof

There are also a variety of topologies, but these agree up to homotopy [8].

For a discrete group G, these various constructions are realizations of the various standard complexes for computing $H_*(G)$. Indeed $B\mathcal{G}$ (or $\mathcal{B}\mathcal{G}$ or $\mathbb{B}G$) is a $K(G,1)$. For any topological group G, the construction $\mathcal{B}\mathcal{G}$ is a realization of the bar construction $\bar{B}$ in homological algebra [3] in the sense that if G is a CW-complex with cellular multiplication (e.g. discrete) then, using cellular chains [5, 9],

$$C_*\mathcal{B}\mathcal{G} \simeq \bar{B}C_*G \quad .$$

MacLane has shown in general that B can be regarded as the tensor product of two functors [4].

In the case of $G = GL_q(\mathbb{F})$, $\mathbb{F} = R$, C or H to recognize $BGL_q(\mathbb{F})$ as the usual Grassmanian up to homotopy, it is perhaps easiest to use a homotopy characterization in terms of a universal bundle [6] or in terms of classification of G-structures as we do in Appendix C.

References

1. A. Dold, Halbexakte Homotopiefunktoren, Lecture Notes in Math., no. 12, Springer-Verlag, Berlin and New York, 1966.

2. A. Dold and R. Lashof, Principal quasi-fibrations and fibre homotopy equivalence of bundles, Illinois J. Math. 3 (1959), 285-305.

3. S. MacLane, Homology, Die Grundlehren der math. Wissenschaften, Band 114, Academic Press, New York; Springer-Verlag, Berlin, 1963.

4. ______, Milgram's classifying space as a tensor product of functors, Steenrod Conference, Lecture Notes in Math., no.168, Springer-Verlag, Berlin and New York.

5. R. Milgram, The bar construction and abelian H-spaces, Illinois J. Math. 11 (1967), 242-250.

6. J.Milnor, Construction of universal bundles. II, Ann. of Math. (2) 63 (1956), 430-436.

7. J.D. Stasheff, A classification theorem for fibre spaces, Topology 2 (1963), 239-246.

8. __________, Associated fibre spaces, Michigan Math. J. 15 (1968), 457-470.

9. __________, H-spaces and classifying spaces, Proc. Symp. Pure Math. 22, AMS, 1971.

10. __________, Homotopy associativity of H-spaces. I , II, Trans. Amer. Math. Soc. 108 (1963), 275-312.

Appendix C - Classification via $\mathcal{B}\Gamma_q$

by J. Stasheff

The approach to classification in § 8 uses the classifying space BX_U associated to an open cover $U = \{U_\alpha\}$. We can construct a space $\mathcal{B}U$ without passing through a category as follows: Order the index set A and let σ now denote a subset $\{ i_0 < i_1 < \dots < i_n\}$. Define $\mathcal{B}U$ as the quotient of $\coprod \Delta_\sigma \times U_\sigma$ by the equivalence relation: if $t_{i_k} = 0$,

$$(t_{i_0},\dots,t_{i_n}, x \in U_\sigma) \sim (\dots,\hat{t}_{i_k},\dots, x \in U_{\sigma'}) \text{ where } \sigma' = (i_0,\dots,\hat{i}_k,\dots).$$

There is an obvious map $r: \mathcal{B}U \to X$ induced by $\Delta_\sigma \times U_\sigma \to U_\sigma$. If U has a subordinate locally finite partition of unity $\{\lambda_\alpha\}$, we can construct $\lambda : X \to \mathcal{B}U$ by choosing for each x, $\sigma(x) = \{\alpha | \lambda_\alpha(x) \neq 0\}$ and then defining $\lambda(x) = (\lambda_{\alpha_0}(x),\dots,\lambda_{\alpha_n}(x), x \in U_{\sigma(x)})$. Clearly $r\lambda = 1$ and it is easy to construct a linear homotopy: $\lambda r \simeq 1$.

The construction and λ look as follows:

By contrast, Segal avoids choosing an ordering essentially by considering the first barycentric subdivision (which has a canonical local ordering). In the second case above this looks like:

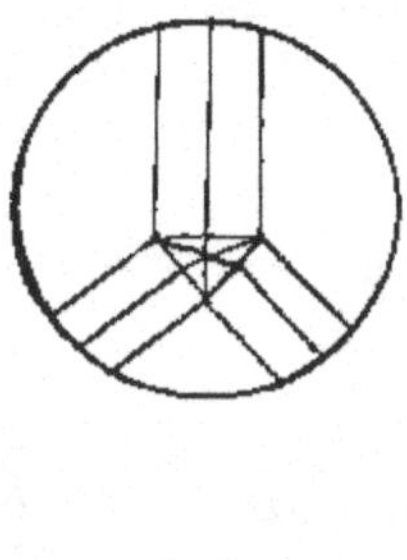

As mentioned in §8, a cocycle $\{U, \gamma_\alpha, \gamma_{\alpha\beta}\}$ induces $BX_U \to B\Gamma_q$. The same holds for $\mathcal{B}U$ if the covering is indexed by the non-negative integers. This is the case for any numerable cover, i.e. one with a subordinate locally finite partition of unity. The definition is even simpler:

$$\Delta_\sigma \times U_\sigma \to \Delta_\sigma \times A_n\Gamma_q$$

is induced by

$$x \longmapsto (\gamma_{i_1 i_0}(x), \ldots, \gamma_{i_n i_{n-1}}(x)) .$$

(For $n = 0$, we use $\gamma_i : U_i \to R^q$.) The cocycle conditions a) and b) are exactly the conditions needed to respect the identifications.

Let us use $\mathcal{B}U$ to prove the classification theorem.

Theorem D. There is a natural 1-1 correspondence between

$$\Gamma_q(X) \text{ and } [X, \mathcal{B}\Gamma_q].$$

We will prove this in such a way that the proof carries over, verbatim, to any topological category C and naturally so with respect to functors $\mathcal{F}: C \to C^1$. In particular, the classification will be compatible with the differential $d: \Gamma_q \to GL_q$.

We have already seen how a cocycle on a numerable cover $\{U_i\}$ induces $\mathcal{B}U \to \mathcal{B}\Gamma_q$. If two cocycles $\{U, \gamma_i, \gamma_{ij}\}$ and $\{V, v_k, v_{k\ell}\}$ are equivalent, order the indices of $U \cup V$ by decreeing those of U to be less than those of V. If the partition of unity for U gives $\lambda : X \to \mathcal{B}U$ and for V we have $\mu : X \to \mathcal{B}V$, use $t\lambda + (1-t)\mu$ to obtain $X \times I \to \mathcal{B}(U \cup V)$ which restricts to λ at one end and μ at the other. The total cocycle assumed on $U\ V$ induces $\mathcal{B}(U \cup V) \to \mathcal{B}\Gamma_q$ and shows we have a well defined map:

$$\psi : H^1(X; \Gamma_q) \to [X, \mathcal{B}\Gamma_q].$$

The relation of homotopy on cocycles induces a homotopy between corresponding classifying maps so, ψ factors:

$$\psi: \Gamma_q(X) \to [X, \mathcal{B}\Gamma_q].$$

We will prove ψ is a bijection by constructing a universal cocycle on $\mathcal{B}\Gamma_q$. We specify the covering in terms of the functions $t_i:\Delta_\sigma \to I$ if $i \in \sigma$, extended to be 0 on Δ_σ if $i \notin \sigma$. Notice these induce well-defined continuous maps $t_i:\mathcal{B}\Gamma_q \to I$. We set $U_i = t_i^{-1}(0,1]$. Trivially the t_i are a partition of unity, but only point finite in general, not locally finite. Following [2 or 3] we shrink gradually to obtain a locally finite partition of unity subordinate to $\{ U_i \} = U$.

For $x \in \mathcal{B}\Gamma_q$, let $W_i(s,x) = \max(0,t_i(x) - s \sum_{j< i} t_j(x))$ and let $v_i(s,x) = W_i(s,x)/ \sum_0^\infty W_j(s,x)$. Note the division is in fact by a finite sum of non-zero terms, hence $v_i:I \times \mathcal{B}\Gamma_q \to I$ is continuous. For $s = 0$, we have $v_i(0,\ ) = t_i$ while at $s = 1$ we have a locally finite partition of unity since, if $\sum_0^n t_j(X) = 1$, then there exists a neighborhood N of x such that $\sum_0^n t_j(y) > 1/2$ for $y \in N$ and hence $v_i(i,y) \leq \max(0,a)$ where $a < 0$ for $i > n$.

Now we are ready to construct our universal example $\gamma \in H^1(\mathcal{B}\Gamma_q;\Gamma_q)$. Define $\gamma_i:U_i \to R^q$ by

$$\gamma_i(t_{i_0},\dots,t_{i_n},f_1,\dots,f_n) = \text{source } f_{i+1} \quad \text{for } i < n$$

$$= \text{target } f_i \quad \text{for } i > 0$$

except that for $n = 0$ let $\gamma_0(1,x) = x \in R^q$.

Similarly define, for $i > j$,

$$\gamma_{ij}(t_{i_0},\ldots,t_{i_n},f_1,\ldots,f_n) = f_i \circ \ldots \circ f_{j+1} .$$

The condition $\gamma_{ij}\gamma_{jk} = \gamma_{ik}$ is immediate if we define $\gamma_{ij} = \gamma_{ji}^{-1}$ for $i < j$ and γ_{ii} to be the appropriate identity. Let γ denote the class of $\{ U ,\gamma_i, \gamma_{ij}\}$ in $H^1(\mathcal{B}\Gamma_q;\Gamma_q)$ or in the set of structures $\Gamma_q(\mathcal{B}\Gamma_q)$.

Given $f: X \to \mathcal{B}\Gamma_q$, we have $f^*(\gamma)$ in $\Gamma_q(X)$ and can apply the natural transformation ψ . To see that $\psi(f^*\gamma) = [f]$, we look directly at the definition. The class $f^*(\gamma)$ is represented on the covering $\{f^{-1}(U_i)\}$ by $\gamma_i \circ f:f^{-1}(U_i)\to R^q$ and $\gamma_{ij}\circ f:f^{-1}(U_i)\cap f^{-1}(U_j) \to \Gamma_q$ As a partition of unity for $\{f^{-1}(U_i)\} = U$, we choose $u_i(x) = v_i(1,f(x))$

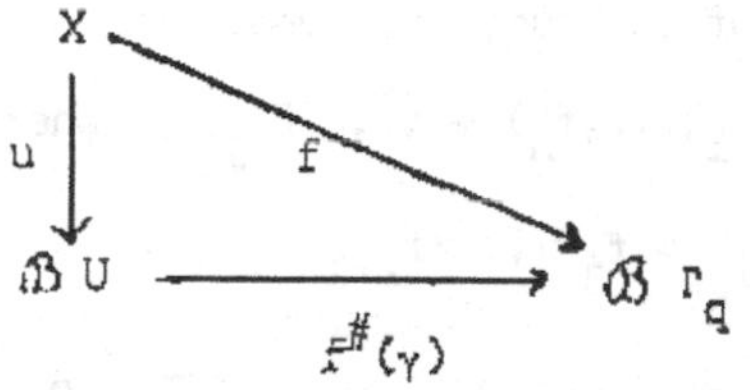

We have, for $x \in U_{\sigma(x)}$, that $f(x)$ can be represented as $(t_{i_0} f(x),\ldots,f_1(x),\ldots,f_n(x))$, so we have $x \to (u_{i_0}(x),\ldots,x) \longrightarrow (u_{i_0}(x),\ldots, \gamma_{i_1 i_0} f(x),\ldots)$ where $\gamma_{i_{j+1} i_j} f(x) = f_{j+1}(x)$. The homotopies $v_i(s,f(x))$ thus define a homotopy from f to $\mathcal{B}f^\#(\gamma) \circ u$. This shows $\psi:H^1(X;\Gamma_q) \to [X, \mathcal{B}\Gamma_q]$ is onto.

On the other hand, let $v = \{V, v_i,v_{ij}\}$ be a cocycle representative of $H^1(X; \Gamma_q)$. If $\lambda_i:V_i \to I$ is a subordinate locally finite partition of unity, we have $((\mathcal{B}v)\circ \lambda)^{-1}(U_i) = \{x|\lambda_i(x) \neq 0\}$, which is a subset of V_i. On that subset, the pull back of the universal γ_i

$H^1(X;C)$ or $C(X)$ signifies unless only invertible morphisms in C are used. However our universal structure involved γ_{ij} only for $i > j$, so our proof does go through in this generality.

Theorem E. Let C be a topological category and X a paracompact space. There is a 1-1 correspondence between $\mathcal{C}(X)$ and $[X, \mathcal{B}C]$.

Indeed, Milnor has pointed out that $\mathcal{B}C$ can be described as "the space of cocycles with values in C." That is, let $\mathcal{B}C$ as a set consist of pairs $(\vec{t}, \{f_{ij}\})$ where $\vec{t} = (t_0, \ldots, t_i, \ldots)$ is a sequence of numbers $t_i \geq 0$, almost all 0 such that $\Sigma t_i = 1$ while $f_{ij} \in \text{Mor}\, C$ ($\text{Ob}C$ if $i = j$) runs over all pairs $i \geq j$ such that $t_i \neq 0$, $t_j \neq 0$ satisfying the relations $f_{ij} \in \text{Mor}(f_{jj}, f_{ii})$ and $f_{ij} \circ f_{jk} = f_{ik}$ for $i > j > k$. To topologize $\mathcal{B}C$ so as to be homeomorphic to $\mathcal{B}C$ as previously defined, use the direct limit of the quotient topologies of the maps $\Delta_\sigma \times A_n C \to \mathcal{B}C$ by $(t_{i_0}, \ldots, f_1, \ldots, f_n) \to (\vec{t}, \{f_{ij}\})$ where $\vec{t}$ is zero except for t_{i_j} and $f_{i_j i_k} = f_j \circ \ldots \circ f_{k+1}$.

Notice the proof is functorial in C, i.e. if $\mathcal{F}: C \to D$ is a continuous functor between topological categories, it induces $\mathcal{B}C \to \mathcal{B}D$ and we have a commutative diagram

$$\begin{array}{ccc} C(X) & \leftrightarrow & [X, \mathcal{B}C] \\ \downarrow & & \downarrow \\ D(X) & \leftrightarrow & [X, \mathcal{B}D] \end{array}$$

In particular this holds for the differential $d: \Gamma_q \to GL_q$ which is a homomorphism (= functor).

If we specialize to a group, we can say more.

Theorem F. Let G be a topological group. For paracompact X, there are natural 1-1 correspondences between the set of equivalence classes of G-bundles over X and $H^1(X;G)$ and $[X, \mathcal{B}\mathcal{G}]$.

The first correspondence is well-known: a bundle can be defined over X from the disjoint union $\coprod U_\alpha \times G$ by identification over is given as follows: $\lambda_i(x)$ appears as some barycentric coordinate of $\mathcal{B}\, v\circ\lambda(x)$ and $v_{ij}(x)$ for some j appears as the corresponding Γ_q-coordinate so we have

$$x \longmapsto (\ldots, \lambda_i(x), \ldots, v_{ij}(x), \ldots) \overset{\gamma_i}{\longmapsto} \text{target } v_{ij} = v_i(x)$$

and similarly the pull back of the universal γ_{ij} agrees with v_{ij} where $\lambda_i(x) \neq 0 \neq \lambda_j(x)$. Thus if $h\colon X \times I \to \mathcal{B}\Gamma_q$ is a homotopy from f_0 to f_1, then $h^*(\gamma)$ restricts as desired to $f_0^*(\gamma)$ and $f_1^*(\gamma)$, showing $\psi : \Gamma_q(X) \to [X, \mathcal{B}\Gamma_q]$ is one-to-one.

Our proofs works in much greater generality. The discussion of Γ_q-structures or a footnote in Segal [4] indicates how to define $H^1(X;C)$ for any topological category C. In terms of an open covering $U = \{U_\alpha\}$, over an ordered index set, a cocycle consists of continuous maps

$$\gamma_\alpha : U_\alpha \to \mathrm{Ob}C$$
$$\gamma_{\alpha\beta} : U_\alpha \cap U_\beta \to \mathrm{Mor}\, C \quad \text{for } \alpha > \beta$$

such that a) $\gamma_{\alpha\beta}(x) \in \mathrm{Hom}(\gamma_\beta(x), \gamma_\alpha(x))$ and b) $\gamma_{\alpha\beta}\gamma_{\beta\delta} = \gamma_{\alpha\delta}$ on $U_\alpha \cap U_\beta \cap U_\delta$ for $\alpha > \beta > \delta$. Two such cocycles $\{U, \gamma_\alpha, \gamma_{\alpha\beta}\}$ and $\{V, \nu_\alpha, \nu_{\alpha\beta}\}$ are cohomologous if they are restrictions of a cocycle on $U \cup V$ with either the indices of U preceding those of V or vice versa. The equivalence classes form $H^1(X;C)$. The set $C(X)$ of C-structures on X is the quotient of $H^1(X;C)$ by the homotopy relation as above. Since our definition depends on the ordering of the covering, it is far from clear what

$U_\alpha \cap U_\beta$ via $\gamma_{\alpha\beta}$. The second correspondence implies one between $H^1(X;G)$ and $G(X)$. That is $H^1(X;G)$ already satisfies the homotopy axiom in this case. This in turn follows from the basic fact of bundle theory: A bundle $p: E \to X \times I$ is equivalent to $p_t \times 1: E_t \times I \to X \times I$ where $p_t = p|p^{-1}(X \times t)$. The example of §7 shows this breaks down for Γ_q-structures even if X is an interval. It is the comparison of the γ_α-part of the cocycle that fails here but not for groups where there is only one object and the γ_α are unique.

The strong result in terms of bundles suggests one further extension of this type of classification, namely to fibrations in the sense of the covering homotopy property rather than local triviality. Here the homotopy is built in, but we lack strict inverses since the appropriate notion of equivalence is fibre homotopy equivalence. Indeed, Dold shows local triviality is often present up to homotopy.

Theorem [2]. Let X be paracompact, weakly locally contractible. Then $p:E \to X$ has the WCHP if and only if the map p is locally fibre homotopy trivial, i.e. for some covering $\{U_\alpha\}$, there exist mutually inverse fibre homotopy equivalences

$$\begin{array}{ccc} p^{-1}(U_\alpha) & \underset{k_\alpha}{\overset{h_\alpha}{\rightleftarrows}} & U_\alpha \times F \\ & \searrow \quad \swarrow & \\ & U_\alpha & \end{array}$$

If we attempt to define a cocycle as before with $H(F) =$ {homotopy equivalence $F \to F$} taking the role of G, we would obtain $(x, \gamma_{\alpha\beta}(x)y) = h_\alpha k_\beta(x,y)$ for $x \in U_\alpha \cap U_\beta$, but the cocycle condition breaks down for we have $h_\alpha k_\beta h_\beta k_\delta \simeq h_\alpha k_\delta$. It turns out that a specific homotopy $\gamma_{\alpha\beta\delta}: I \times U_\alpha \cap U_\beta \cap U_\delta \to H(F)$ is relevant to classifying the fibration. Wirth [7] has carried out a full program along these lines and finds higher homotopies $I^{n-1} \times U_\sigma \to H(F)$ are also important. One way of summarizing his result is the following: We have seen that the cocylce condition corresponds to an associated map being a functor or, in pseudogroup terminology, a homomorphism. With topological categories or topological pseudogroups, it makes sense to talk of functors "up to strong homotopy" or s(trongly) h(omotopy) m(ultiplicative) maps [6]. Associated to these conditions, we can define a "cocycle up to strong homotopy." Similarly the equivalence or cobounding relation can with effort be generalized up to homotopy so as to define $H^1_\pi(X,G)$ as a limit over all covers of X.

Theorem G. [Wirth, 7]. There is a 1-1 correspondence between fibre homotopy equivalence classes of fibrations (with the WCHP) over paracompact, weakly locally contractible X with fibres of the homotopy type of F and the set $H^1_\pi(X; H(F))$.

Since shm maps induce maps of classifying spaces, so do functors up to strong homotopy. Thus either of the above sets is in 1-1 correspondence with $[X, \mathcal{B}H(F)]$, as was known for the first mentioned set [1, 5].

References

1. A. Dold, Halbexakte Homotopiefunktorem, Lecture Notes in Math, No. 12, Springer-Verlag, Berlin and New York, 1966.

2. ______, Partitions of unity in the theory of fibrations, Ann. of Math. (2) 78 (1963), 223-228.

3. D. Husemoller, Fibre Bundles, McGraw-Hill, New York, 1966.

4. G. Segal, Classifying spaces and spectral sequences, Inst. Hautes Études Sci. Publ. Math., No. 34 (1968), 105-112.

5. J.D. Stasheff, A classification theorem for fibre spaces, Topology 2 (1963), 239-246.

6. ______, H-spaces from a homotopy point of view, Lecture Notes in Math, 161, Springer-Verlag, Berlin and New York, 1970.

7. J. Wirth, Fibre spaces and the higher homotopy cocycle relations, Thesis, Notre Dame, Ind., 1964.

OPERACIONES COHOMOLOGICAS DE ORDEN SUPERIOR

por Samuel Gitler

con la colaboración de Carlos Ruiz

0. Introducción

Sea π un grupo abeliano, denotamos por $K(\pi,n)$ a cualquier espacio del tipo de homotopía de un CW tal que

$$\pi_n(K(\pi,n)) = \begin{cases} \pi, & q = n \\ 0, & q \neq n. \end{cases}$$

Entonces por el teorema de Hurewicz,

$$H_i(K(\pi,n);\underline{Z}) = 0, \quad i<n;$$

$$H_n(K(\pi,n);\underline{Z}) = \pi_n(K(\pi,n)) = \pi,$$

y tenemos

$$H^n(K(\pi,n);\pi) \cong \operatorname{Hom}(H_n(K(\pi,n);\underline{Z}),\pi) = \operatorname{Hom}(\pi,\pi);$$

el primer isomorfismo por la fórmula de coeficientes universales. Luego al isomorfismo identidad de π corresponde una clase $\gamma_n \in H^n(K(\pi,n);\pi)$ llamada la clase fundamental de $K(\pi,n)$.

Dada $f: X \to K(\pi,n)$, le corresponde una clase $f^*\gamma_n \in H^n(X,\pi)$, que depende sólo de la clase de homotopía de f.

Designemos por $[f]$ a la clase de homotopía de aplicaciones $X \to Y$, se tiene el siguiente teorema:

Teorema 0.1. *Sea X un complejo CW, la correspondencia que asocia a $[f] \in [X,K(\pi,n)]$ la clase $f^*\gamma_n$, es una correspondencia biyectiva.*

Si $x = f^*\gamma_n$ decimos que f "clasifica" a x.

El teorema 0.1 es consecuencia de la teoría de obstrucciones, una demostración puede verse en [10, pp. 6-10] o bien en [14, th. 8.18].

Si X es un espacio con un punto base, que indicamos con $*$, dentamos con $\mathcal{P}X$ al espacio de caminos, es decir, el conjunto de todas las aplicaciones $\alpha:[0,1] \to X$ tales que $\alpha(0) = *$, con la topología compacto-abierta.

Sea $p: \mathcal{P}X \to X$ la aplicación $p(\alpha) = \alpha(1)$, entonces $p^{-1}(*) = \Omega X$ es el <u>espacio</u> <u>de</u> <u>lazos</u> de X.

$\mathcal{P}X$ es un espacio contráctil y

$$\Omega X \longrightarrow \mathcal{P}X \xrightarrow{p} X$$

es una fibración [14, 2.8.8], llamada la fibración de caminos de X.

En particular, si $X = K(\pi,n)$, de la sucesión exacta de homotopía se deduce que $\Omega K(\pi,n) = K(\pi,n-1)$.

Una fibración principal es una fibración

$$F \xrightarrow{i} E \xrightarrow{p} B$$

con una acción $\mu: F \times E \longrightarrow E$, fibra a fibra, que convierte a F en un H espacio y con una aplicación

$$\nu: \bar{E} \longrightarrow F,$$

donde

$$\bar{E} = \{(x,y) \in E \times E / p(x) = p(y)\},$$

$$\begin{array}{ccc} \bar{E} & \xrightarrow{\pi_2} & E \\ {\scriptstyle \pi_1}\downarrow & & \downarrow{\scriptstyle p} \\ E & \xrightarrow{p} & B, \end{array}$$

tal que

$$\mu \circ (\nu \times \pi_1) \simeq \pi_2 \text{ [12]},$$

ver [12].
La fibración de caminos es una fibración principal y toda fibración inducida de una fibración principal es también una fibración principal.

<u>Proposición</u> 0.2. <u>Sea</u> $F \longrightarrow E \xrightarrow{p} B$ <u>una fibración principal y</u> $\varphi : X \longrightarrow B$.

<u>Si</u> $\tilde{\varphi}$ <u>y</u> $\tilde{\tilde{\varphi}}$ <u>son dos levantamientos de</u> φ <u>a</u> E, <u>existe</u> $g : X \longrightarrow F$ <u>tal que</u>

$$X \xrightarrow{\Delta} X \times X \xrightarrow{g \times \tilde{\varphi}} F \times E \xrightarrow{\mu} E$$

<u>es homotópica a</u> $\tilde{\tilde{\varphi}}$ [12],[13].

Dada una fibración

$$f \xrightarrow{i} E \xrightarrow{p} B$$

tenemos el siguiente diagrama

$$\begin{array}{ccccccccc}
\longrightarrow H^n(E) & \xrightarrow{i^*} & H^n(F) & \xrightarrow{\delta} & H^{n+1}(E,F) & \xrightarrow{j^*} & H^{n+1}(E) & \longrightarrow \\
 & & & & \uparrow p_1^* & & \uparrow p^* & \\
 & & & & H^{n+1}(B,*) & \underset{k^*}{\xrightarrow{\approx}} & H^{n+1}(B) &
\end{array}$$

donde la fila superior es la sucesión exacta del par (E,F).

Sea $T = \delta^{-1}(\text{Im } p_1^*)$ y sea

$$\tau : T \longrightarrow H^{n+1}(B)/k^*(\ker p_1^*),$$

$$\tau = k^* p_1^{*-1} \delta .$$

τ se llama la <u>trasgresión</u> de la fibración y T el subgrupo de los elementos <u>trasgresivos</u>.

Es fácil verificar lo siguiente:

$$\ker \tau = \operatorname{Im} i^* = \ker \delta$$

$$\operatorname{Im} \tau = \ker p^*/k^*(\ker p_1^*).$$

La relación inversa

$$\sigma^* = \delta^{-1} p_1^* k^{-1} : \ker p^* \longrightarrow H^n(F)/\operatorname{Im} i^*$$

se llama la <u>suspensión</u>.

Entonces τ y σ^* inducen isomorfismos inversos

$$T/\operatorname{Im} i^* \longleftrightarrow \ker p^*/k^*(\ker p_1^*).$$

La trasgresión tiene una definición en términos de la sucesión espectral de la fibración, para esta noción referimos a [7],[14] . Por este camino se demuestra el siguiente teorema debido a Serre [11].

<u>Teorema</u> 0.3. <u>Si</u> B <u>es</u> p-<u>conexo</u> <u>y</u> F <u>es</u> q-<u>conexo</u>, <u>entonces</u> <u>para</u> $n \leq p+q+1$, $T = H^n(F)$; $\ker p_1^* = 0$ <u>y</u> <u>la</u> <u>siguiente</u> <u>sucesión</u> <u>es</u> <u>exacta</u>

$$\longrightarrow H^{n-1}(E) \xrightarrow{i^*} H^{n-1}(F) \xrightarrow{\tau} H^n(B) \xrightarrow{p^*} H^n(E) \xrightarrow{i^*} H^{n}(F).$$

Si X, Y son espacios con un punto base (que denotamos siempre *), definimos

1) la <u>suma</u> <u>reducida</u>

$$X \vee Y = \{(x,*)\} \vee \{(*,y)\} \subset X \times Y,$$

2) el <u>producto</u> <u>reducido</u>

$$X \wedge Y = X \times Y / X \vee Y,$$

en particular,

2 a) la <u>suspensión</u> de X es

$$\Sigma X = X \wedge S^1 = X \times I/\sim$$

donde $\sim$ es la relación que identifica entre sí a todos los puntos de la forma

$$(*,t),(x,0),(x,1).$$

Notar que $S^m \wedge S^n = S^{m+n}$.

3) La <u>junta</u> (join)

$$X*Y = X+X \times I \times Y+Y/\sim$$

donde $\sim$ es la relación que identifica $x = (x,0,y)$ para todo $y \in X$.

<u>Proposición</u> 0.4. $\Sigma(X \wedge X) \simeq X*X$.

Si G es un H-espacio asociativo, existe un espacio <u>clasificante</u> BG con la propiedad de que $[X,BG]$ está en correspondencia biunívoca con las clases de isomorfismo de fibraciones principales sobre X, [9].

Sea μ la multiplicación de G. Entonces por la proposición (0.3) la restricción de $\Sigma\mu$ es equivalente, salvo homotopías con la aplicación $\bar{\mu}: G*G \longrightarrow \Sigma G$, que se obtiene de la construcción de Hopf. Se ve que μ es una G-fibración principal, sea $q:\Sigma G \longrightarrow BG$ su aplicación clasificante. Entonces tenemos una fibración

$$G*G \xrightarrow{\bar{\mu}} \Sigma G \xrightarrow{q} BG,$$

se ve en [2] que q es adjunta de id_G en la correspondencia

$$[G,G] = [G,\Omega BG] \longleftrightarrow [\Sigma G,BG] \quad \text{(ver [14])}$$

En particular para $G = K(\underline{Z}_2,n)$, tenemos

$$(0.5) \qquad K(\underline{Z}_2,n)*K(\underline{Z}_2,n) \xrightarrow{\bar{\mu}} \Sigma K(\underline{Z}_2,n) \xrightarrow{g} K(\underline{Z}_2,n+1)$$

1. Operaciones cohomológicas primarias y el álgebra de Steenrod módulo 2

Una operación cohomológica de tipo (π,n,G,q) es una transformación natural de functores

$$\theta : H^n(\quad;\pi) \longrightarrow H^q(\quad;G);$$

es decir que consiste de un homomorfismo

$$\theta_X : H^n(X;\pi) \longrightarrow H^q(X;G)$$

para cada espacio X, de manera que para cada $f:Y \to X$ continua el siguiente diagrama sea conmutativo:

$$\begin{array}{ccc} H^n(X,\pi) & \xrightarrow{\theta_X} & H^q(X,G) \\ \downarrow f^* & & \downarrow f^* \\ H^n(Y,\pi) & \xrightarrow{\theta_Y} & H^q(Y,G) \end{array}$$

Al conjunto de todas las operaciones cohomológicas de tipo (π,n,G,q) lo designamos $O(\pi,n,G,q)$.

<u>Teorema</u> 1.1. <u>La aplicación que le asocia a cada operación cohomológica</u> $\theta \in O(\pi,n,G,q)$, <u>la clase de cohomología</u> $\theta\gamma_n \in H^q(K(\pi,n);G)$, <u>es biyectiva</u>.

<u>Demostración</u>. Construiremos la aplicación inversa:

dado $u \in H^q(K(\pi,n);G)$ le asociamos una operación cohomológica θ_u como sigue. Si $x \in H^n(X,\pi)$, según (Teorema 0.1) x y u están representadas por aplicaciones φ_x y φ_u respectivamente.

$$X \xrightarrow{\varphi_x} K(\pi,n) \xrightarrow{\varphi_u} K(G,q)$$

entonces definimos $\theta_u(x)$ como la clase de cohomología (perteneciente a $H^q(X,G)$) clasificada por $\varphi_u \circ \varphi_x$. □

Definimos: $S^q = \lim \text{proy}_n H^{q+n}(K(\pi,n);G)$, el límite inverso, tomado según las suspensiones en las fibraciones de lazos de $K(\pi,n)$:

$$G^*:H^{q+n+1}(K(\pi,n+1);G) \longrightarrow H^{q+n}(K(\pi,n);G),$$

que según (0.3) son isomorfismos para $q \leq 2n$, $S^* = \left\{S^q\right\}$.

Entonces, $S^q(\pi,G)$ es un grupo y sus elementos son las operaciones estables de tipo (π,G,q); es decir que para $\theta \in S^q(\pi,G)$ y $x \in H^n(X;\pi)$ tenemos $\theta x \in H^{n+q}(X;G)$.

En particular, si $\pi = G = \underline{Z}_2$, $S^*(\underline{Z}_2,\underline{Z}_2)$ es un álgebra bajo la composición, llamada el álgebra de Steenrod módulo 2 y denotada $\mathcal{A}(2)$ o simplemente $\mathcal{A}$.

Ahora describiremos un sistema de generadores de $\mathcal{A}$.

Pongamos $K_n = K(\underline{Z}_2,n)$; $\gamma_n \in H^*(K_n;\underline{Z}_2)$ genera un álgebra de polinomios ya que si $x \in H^1(RP^\infty;\underline{Z}_2)$ es el generador de $H^*(RP^\infty,\underline{Z}_2) = \underline{Z}_2[x]$ y $f:RP^\infty \longrightarrow K_n$ clasifica a x^n, es decir, $f^*\gamma_n = x^n$, entonces $f^*(\gamma_n) = x^{nt} \neq 0$.

Consideremos la fibración (0.5)

$$K_n * K_n \xrightarrow{\bar{\mu}} \Sigma K_n \xrightarrow{g} K_{n+1},$$

como γ_n^2 es primitivo $\bar{\mu}^*\Sigma^*\gamma_n^2 = 0$; luego, en la sucesión exacta de

Serre, $\Sigma^*\gamma_n^2$ proviene de K_{n+1}, digamos

$$\Sigma^*\gamma_n^2 = g^*\alpha_{n+1},$$

con α_{n+1} en $H^{2n+1}(K_{n+1})$ único.

Ahora bien, que g sea adjunta de la identidad significa que el siguiente diagrama, donde σ^* es la suspensión de la fibración de lazos de K_{n+1} y Σ^* corresponde a la suspensión geométrica, es conmutativo

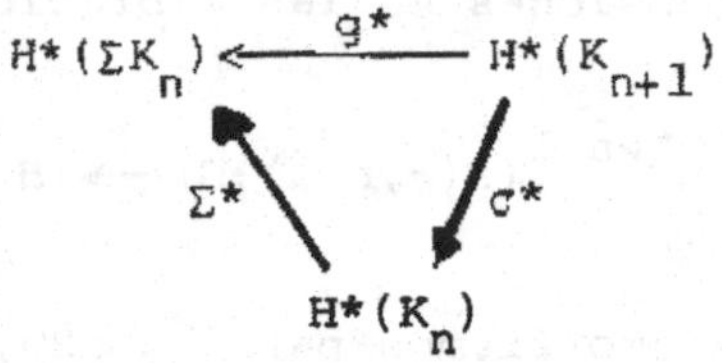

entonces $\sigma^*\alpha_{n+1} = \gamma_n^2$.

Supongamos tener elementos

$$\alpha_i \in H^{i+n}(K_i, \underline{Z}_2), \quad n+1 \leq i < j$$

tales que

$$\sigma^*\alpha_{i+1} = \alpha_i$$

en la fibración de lazos de K_{i+1}; como K_{j+1} es j-conexo, usando la sucesión exacta de Serre se deduce que existe un único elemento elemento $\alpha_{j+1} \in H^{j+n+1}(K_{j+1}, \underline{Z}_2)$ tal que $\sigma^*\alpha_{j+1} = \alpha_j$.

Entonces $(\alpha_i)_{i \geq n}$ es un elemento de G_n que denotamos Sq^n y llamamos el n-simo <u>cuadrado de Steenrod</u>.

<u>Teorema</u> 1.2. <u>Los cuadrados de Steenrod verifican las siguientes propiedades</u>:

1. $Sq^n: H^i(X) \longrightarrow H^{i+n}(X)$

2. *Si* $x \in H^n(X)$, $Sq^n x = x^2$

3. *Si* $x \in H^i(X)$, $i < n$, $Sq^n x = 0$

4. $Sq^0 = 1 : H^i(X) \longrightarrow H^i(X)$

5. $Sq^1 = H^i(X) \longrightarrow H^{i+1}(X)$ *coincide con el homomorfismo de Bockstein asociado a la sucesión*

$$0 \longrightarrow \underline{Z}_2 \longrightarrow \underline{Z}_4 \longrightarrow \underline{Z}_2 \longrightarrow 0$$

6. *Fórmula de Cartan*:

$$Sq^n(xy) = \Sigma_{i=0}^{n} Sq^i x Sq^{n-i} y$$

7. *Relaciones de Adem*: Si $a < 2b$

$$Sq^a Sq^b = \Sigma_{i=0}^{[a/2]} \binom{b-i-1}{a-2i} Sq^{a+b-i} Sq^i$$

donde $\binom{m}{k}$ *representa al número combinatorio, tomando módulo* 2 *y* [n] *indica parte entera.*

Para esto y lo que sigue ver referencias [10] y [15].

Teorema 1.3. *El álgebra de Steenrod* A *es el álgebra generada por los cuadrados* Sq^i *sujetos a las relaciones de Adem.*

Digamos que una sucesión de números enteros $I = (i_1, \ldots, i_2)$ es admisible si $i_j \geq 2 i_{j+1}$.

Pongamos $Sq^I = Sq^{i_1} \ldots Sq^{i_r}$.

Teorema 1.4. *Los* Sq^I, I *admisible son una* $\underline{Z}_2$*-base de* $\mathfrak{a}$.

2. Operaciones secundarias

Sea $\sum_{k=1}^{m} \alpha_k \beta_k = 0$ una relación homogénea de grado $n+1$ en $\mathfrak{a}$.

Para simplificar la notación escribiremos $\alpha = (\alpha_1, \ldots, \alpha_m)$,

$\beta = (\beta_1, \dots, \beta_m)$; entonces la relación se escribe $\alpha\beta = 0$.

Ponemos $\dim \beta_k = t_k$.

Definimos:

$$N^q(\beta, X) = \{x \in H^q(X); \beta_k(x) = 0,\ k = 1, \dots, m\}$$
$$I^{q+n}(\alpha, X) = \Sigma\alpha_k(H^{q+n-\dim \alpha_k}(X)).$$

Una <u>operación secundaria</u> Φ asociada con la relación $\alpha \cdot \beta = 0$ es una transformación (para cada q)

$$\Phi : N^q(\beta,\ ) \longrightarrow H^{q+n}(\ )/I^{q+n}(\alpha,\ ) \tag{2.1}$$

que verifica los siguientes axiomas:

<u>Axioma</u> 1 (Naturalidad). Si $f: Y \longrightarrow X$ es una función continua y $x \in N^q(\beta, X)$, entonces $f^*X \in N^q(\beta, Y)$ y $f^* \Phi(x) = \Phi(f^*x)$, módulo $I^{q+n}(\alpha, Y)$.

Para el axioma 2 hace falta alguna notación:

Si (X,Y) es un par, $Y \xrightarrow{i} X \xrightarrow{j} (X,Y)$, las inclusiones $v \in H^q(X,Y)$, $j^*v = x$ y $j^*v \in N^q(\beta, X)$, para cada k tenemos un diagrama:

$$\begin{array}{ccccccc}
\xrightarrow{i^*} & H^{q-1}(Y) & \xrightarrow{\delta} & H^q(X,Y) & \xrightarrow{j^*} & H^q(X) & \\
 & \downarrow{\scriptstyle \beta_k} & & \downarrow{\scriptstyle \beta_k} & & \downarrow{\scriptstyle \beta_k} & \\
\longrightarrow & H^{q-1+t_k}(\ ) & \longrightarrow & H^{q+t_k}(X,Y) & \longrightarrow & H^{q+t_k}(X) & \longrightarrow
\end{array}$$

Se deduce que existen elementos $w_k \in H^{q+t_k-1}(Y)$ tales que $\delta w_k = \beta_k v$.

<u>Axioma</u> 2 (Fórmula de Peterson-Stein).

$$i^* \Phi(x) = \Sigma\alpha_k w_k.$$

Una operación secundaria se llama <u>estable</u> si verifica además:

<u>Axioma</u> 3 (Estabilidad). Si $x \in N^q(\beta,X)$ entonces $\Sigma^*x \in N^{q+1}(\beta,\Sigma X)$ y $\Phi(\Sigma^*x) = \Sigma^*\Phi(x)$.

Vamos a construir operaciones secundarias estables por el método del ejemplo universal.

Sea $q > n$ y sea $f_k: K_q \longrightarrow K_{q+t_k}$ la aplicación que clasifica a la clase $\beta_k\gamma_q$ es decir $f_k^*\gamma_{q+t_k} = \beta_k\gamma_q$; las f_k, $k = 1,\ldots,m$ definen $f: K_q \longrightarrow \Pi_{k=1}^m K_{q+t_k}$.

De la fibración de caminos de $\Pi_{k=1}^m K_{q+t_k}$, f induce una fibración

(2.2) $$\Pi K_{q+t_k-1} \longrightarrow E_q \xrightarrow{\Pi} K_q$$

En la fibración de curvas debe ocurrir: $\tau\gamma_{q+t_k-1} = \gamma_{q+t_k}$, luego, por naturalidad, en la fibración (2.1):

$$\tau\gamma_{q+t_k-1} = \beta_k\gamma_q.$$

De $\alpha\cdot\beta = 0$ deducimos que $\Sigma\alpha_k\gamma_{q+t_k-1}$ está en el núcleo de τ, por lo tanto existe $v_q \in H^{q+n}(E_q)$ tal que

$$i^*v_q = \Sigma\alpha_k\gamma_{q+t_k-1}.$$

Se ve que existe una fibración $E_{q+1} \longrightarrow K_{q+1}$ que da $E_q \longrightarrow K_q$ al tomar lazos entonces usando (0.3) para la fibración de caminos de E_{q+1} resulta que existe un único elemento v_{q+1} tal que $\sigma^*v_{q+1} = v_q$. Sea $v = \{v_i\}$. Veamos que v define una operación secundaria asociada con $\alpha\beta = 0$.

Sea $x \in H^q(X)$ tal que $\beta_k x = 0$ $k = 1,\ldots,m$, quiere decir que la composición $X \xrightarrow{\varphi_x} K_q \xrightarrow{f} \Pi K_{q+t_k}$ es homotópica a la constante. Entonces φ_x se levanta a $\tilde{\varphi}_x : X \to E_q$. Definimos $\Phi_v(x) = \{\tilde{\varphi}_x^*(v_q) | \tilde{\varphi}_x$ es un levantamiento de $\varphi_x\}$. Luego Φ_v está definido en $N^q(\beta, X)$, veamos que toma valores en $H^{q+n}(X)/I^{q+n}(\alpha, X)$. Por ser la fibración principal, si $\tilde{\varphi}_x$ y $\tilde{\varphi}'_x$ son dos levantamientos de φ_x existe $g: X \to \Pi K_{q+t_k-1}$ tal que la composición:

$$X \xrightarrow{\Delta} X \times X \xrightarrow{g \times \tilde{\varphi}_x} \Pi_{q+t_k-1} \times E_q \xrightarrow{\mu} E_q$$

es homotópica a $\tilde{\varphi}'_x$.

Ahora:

$$\mu^*(v_q) = 1 \otimes v_q + \Sigma \alpha_k \gamma_{q+t_k-1} \otimes 1,$$

entonces

$$\Delta^*(g \times \varphi_x)^* \mu^*(v_q) = \varphi_x^* v_q + \Sigma \alpha_k g^* \gamma_{q+t_k-1}$$

pero el segundo numerando de la derecha es un elemento arbitrario de $I^{q+n}(\alpha, X)$, o sea los valores de $\Phi_v(x)$ están determinados a menos de $I^{q+n}(\alpha, X)$.

Probaremos ahora que Φ_v satisface los axiomas:

<u>Axioma</u> 1. Sea $f: Y \to X$. Si $\beta_k x = 0$, $f^* \beta_k x = \beta_k f_x^* = 0$, luego $x \in N^q(\beta, X)$ implica $f^*(x) \in N^q(\beta, Y)$.

Si $\tilde{\varphi}_x$ es un levantamiento de φ_x entonces $\tilde{\varphi}_x \circ f$ es un levantamiento de $\varphi_{f_x^*}: Y \to K_q$, luego $(\tilde{\varphi}_x \circ f)^* v_q = f^* \tilde{\varphi}_x^* v_q$ representa a $\Phi_v(f_x^*)$, o sea $f^* \Phi_v(x) = \Phi_v(f_x^*)$.

<u>Axioma</u> 2. En E_q tenemos $\bar{i}^* v_q = \Sigma \alpha_k \gamma_{q+t_k-1}$. Consideremos

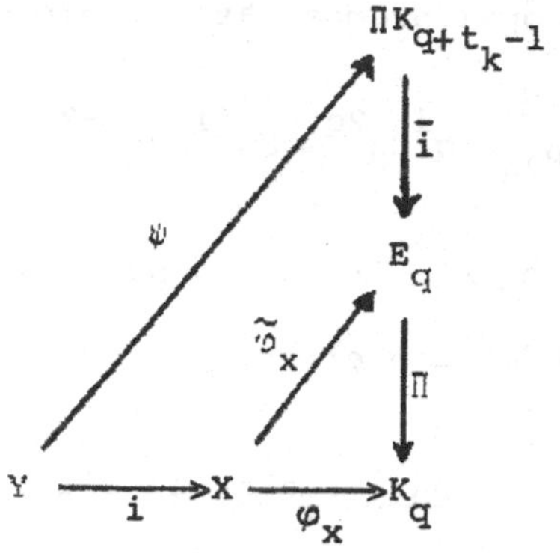

ψ existe puesto que $i*\phi_x^*\gamma_q = i*x = 0$.

Tenemos $i*\widetilde{\phi}_x^*(v_q) = \psi*\bar{i}*v_q = \Sigma\alpha_k\psi*\gamma_{q+t_k-1}$. Por otra parte consideremos el diagrama conmutativo, donde hemos puesto

$F_q = \amalg K_{q+t_k-1}$,

$$\longrightarrow H^{t-1}(Y) \xrightarrow{\delta} H^t(X,Y) \xrightarrow{j*} H^t(X) \xrightarrow{i*} H^t(Y) \longrightarrow$$

$$\uparrow{\psi*} \qquad \uparrow{\rho_x^*} \qquad \uparrow{\widetilde{\phi}_x^*} \qquad \uparrow{\psi*}$$

$$\longrightarrow H^{t-1}(F_q) \xrightarrow{\tau} H^t(K_q) \xrightarrow{\Pi*} H^t(E_q) \xrightarrow{\bar{i}*} H^t(F_q) \longrightarrow$$

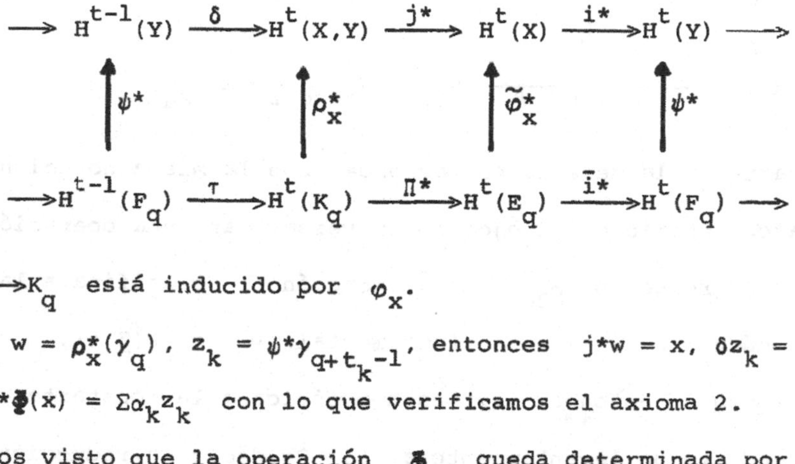

$\rho_x : X/Y \longrightarrow K_q$ está inducido por ϕ_x.

Sea $w = \rho_x^*(\gamma_q)$, $z_k = \psi*\gamma_{q+t_k-1}$, entonces $j*w = x$, $\delta z_k = \beta_k w$, luego $i*\Phi(x) = \Sigma\alpha_k z_k$ con lo que verificamos el axioma 2.

Hemos visto que la operación Φ_v queda determinada por la elección de v_q, si elegimos otro elemento v_q', de la sucesión exacta se deduce que $v_q - v_q'$ proviene de $H^{q+n}(K_q)$ digamos $v_q - v_q' = \Pi*\theta$, por lo tanto para un mismo levantamiento $\widetilde{\phi}_x$ de ϕ_x, los valores de Φ_v y $\Phi_{v'}$ difieren en θ, es decir, en una operación primaria estable.

Veamos ahora una familia interesante de operaciones secundarias.

Para $q \geq 2$ consideremos las relaciones

$$\rho_{2q}: Sq^1 Sq^{2q} + Sq^{01} Sq^{2q-2} + Sq^{2q} Sq^1 = 0$$

(donde hemos puesto $Sq^{01} = Sq^3 + Sq^2 Sq^1$) y sean Φ_{2q} operaciones secundarias asociadas con ellas.

<u>Teorema</u> 2.3. a) <u>Para cualquier clase entera</u> $x \in H^t(X)$, $\Phi_{2q}(x)$ <u>está definida para</u> $t \leq 2q-3$. b) <u>Existe una operación</u> Φ_{2q} <u>tal que si</u> $t = 2q-3$, $\Phi_{2q}(x) = 0$.

<u>Demostración</u>. Consideremos el diagrama:

$$\begin{array}{ccccc} \bar{E}_t & \xrightarrow{r'} & E_t & & \\ \downarrow{\scriptstyle \pi'} & & \downarrow{\scriptstyle \pi} & & \\ K(\underline{Z},t) & \xrightarrow[r]{} & K_t & \xrightarrow[f]{} & K_{t+1} \times K_{t+2q-2} \times K_{t+2q} \end{array}$$

La parte de la derecha corresponde, con la misma notación que en los parrafos anteriores al ejemplo universal para una operación Φ_{2q} asociada a la relación ρ_{2q}. La aplicación r clasifica a la reducción módulo 2 de la clase fundamental u de $K(\underline{Z},t)$.

Si $t \leq 2q-3$ entonces fr es homotópica a la constante y por lo tanto r admite un levantamiento a $\tilde{r}$. Es decir la reducción módulo 2 de u pertenece a $N^t(\beta, K(\underline{Z},t))$. Esto prueba a).

Veamos b), tenemos $\bar{E}_t = K(\underline{Z},t) \times K_t \times K_{t+2q-3} \times K_{t+2q-1}$ como $i^* v_q = Sq^1 \gamma_{t+2q-1} + Sq^{01} \gamma_{t+2q-3} + \pi'^* b_{t+2q}$, con $b \in H^{t+2q}(K_t)$. Pongamos $t = 2q-3$ y consideremos la fibración (ver 0.5),

$$\bar{E}_{2q-3} * \bar{E}_{2q-3} \xrightarrow{\bar{\mu}} \Sigma \bar{E}_{2q-3} \longrightarrow \bar{E}_{2q-2};$$

es fácil ver que $\bar{\mu}^*(\Sigma^*\gamma_{49-6}) = u_{2q-3}*u_{2q-3}$. Ahora $r'^*(v_{2q-3}) = Sq^1\gamma_{4q-4}+Sq^{01}\gamma_{4q-6}+\pi'^*b_{4q-3}$ es primitivo en $\bar{E}_{2q-3}$ por estar en la imagen de la suspensión.

Pero $\bar{\mu}^*\Sigma^*Sq^{01}\gamma_{4q-6} = Sq^3u*u+u*Sq^3u$ por lo que

$$r'^*(v_{2q-3}) = Sq^1\gamma_{49-4}+Sq^{01}\gamma_{4q-6}+\pi^*(Sq^3u\smile u)+\pi'^*b'_{4q-3},$$

con b'_{4q-3} primitivo.

Ahora bien si b'_{4q-3} es primitivo entonces es suspensión de un elemento $b'_{4q-2}\in H^{4q-2}(K_{2q-2})$ pero en los complejos de Eilenberg MacLane σ^* sólo manda operaciones estables en estables luego b'_{4q-2} es estable y b'_{4q-3} también.

Entonces cambiando v_{2q-3} por $v_{2q-3}-b'_{4q-3}$ obtenemos

$$\Phi_{2q}(u) = Sq^3u\smile u \quad \text{módulo}$$

$$Sq^1H^{2q-4}(K(\underline{Z},2q-3))+Sq^{01}H^{2q-6}(K(\underline{Z},2q-3))$$

pero $Sq^1(sq^2u\smile u) = Sq^3u\smile u$ luego $\Phi(u) = 0$ para u, la clase fundamental de $K(\underline{Z},2q-3)$.□

Veamos una manera de calcular estas operaciones. Para las nociones relativas a los haces lineales (fibrados vectoriales) remitimos a [4]. X^ξ es el espacio de Thom y c_q las clases de Chern.

Entre todos los Φ_{4q} elegimos la que toma el valor 0 en dimensiones bajas, a la que llamaremos la Φ_{4q} normalizada.

<u>Teorema</u> 2.4. <u>Sea ξ un fibrado vectorial complejo de dimensión compleja</u> n, <u>sobre un espacio</u> X. <u>Sea</u> $v\in H^{2n}(X^\xi;\underline{Z})$ <u>su clase de Thom. Supongamos que</u> $c_{2q}(\xi) = 2x$ <u>y</u> $c_{2q-1}(\xi) = 2y$. <u>Entonces</u>

$\Phi_{4q}(v)$ <u>está definida y</u> $\Phi_{4q}(v) = vx_{2q}+Sq^2(vx_{2q-2})$.

La demostración de (2.4) se encuentra en [4].

Como consecuencia tenemos:

$$\Phi_{2^r}(\omega^{2^r}) = \omega^{2^r+2^{r-1}} .$$

Observese que en este caso como $Sq^j(\omega^{2^r}) = 0$ para $j \leq 2^r$, cualquier operación secundaria asociada con ρ_{2r} tiene el mismo valor en ω^{2^r}.

Como aplicación veamos que no existe $S^{31} \xrightarrow{f} S^{16}$ con invariante de Hopf 1. Si lo hubiere, obtendríamos $Sq^{16}: H^{16}(S^{16} \cup_f e^{32}) \to H^{32}(S^{16} \cup_f e^{32})$ es no trivial. Pero ahora si consideramos el ejemplo

$$\begin{array}{l} E_q \\ \downarrow \\ K_q \xrightarrow{g} K_{q+1} \times K_{q+2} \times K_{q+4} \times K_{2+8} \end{array}$$

con $g(\gamma_{q+2^i}) = Sq^{2^i}\gamma_q$ y $i = 0,1,2,3$, entonces en E_q existen clases v_1 asociada con $Sq^1Sq^1 = 0$, v_3 asociada con $Sq^2Sq^2+Sq^3Sq^1 = 0$, v_4 asociada con $Sq^1Sq^4+Sq^2Sq^1Sq^2+Sq^4Sq^1 = 0$, v_7 asociada con $Sq^4Sq^4+Sq^7Sq^1+Sq^6Sq^2 = 0$, v_8 con $Sq^1Sq^8+Sq^8Sq^1+Sq^2Sq^1Sq^6 = 0$, v_9 con $Sq^2Sq^8+Sq^8Sq^2+Sq^4Sq^6+Sq^9Sq^1 = 0$, y v_{15} asociada con $Sq^8Sq^8+Sq^{15}Sq^1+Sq^{13}Sq^2+Sq^{11}Sq^4 = 0$ y estas clases satisfacen una relación del tipo:

(2.5) $$Sq^1v_{15}+\theta_7v_9+\theta_8v_8+\theta_9v_7+\theta_{12}v_4+\theta_{13}v_3+\theta_{15}v_1 = \lambda Sq^{16}$$

con $\theta_8 = Sq^8+\ldots$. Si nos fijamos en $\omega^8 \in H^{16}(CP^\infty)$, $\phi_8(\omega^8) = \omega^{12}$ y $Sq^8\phi_8(\omega^8) = \omega^{16}$, luego $\lambda \neq 0$. Ahora en $S^{16} \cup e^{32}$ si $Sq^{16} \neq 0$, por la relación (2.5) alguna de las operaciones Φ_k asociadas con v_k es no nula en i_{16}, pero eso es una contradicción.

3. Espectros y torres

Un espectro es una colección de espacios y funciones

$X = \{X_k, f_k\}_{k \in Z}$, $f_k : \Sigma X_k \longrightarrow X_{k+1}$.

Definimos $H^q(\underline{X};G) = \lim \text{proy}\, H^{q+k}(X_k;G)$ y $\Pi_q(\underline{X}) = \lim \text{dir}\, \Pi_{q+k}(X_k)$. $\underline{X}$ es estable si existe una función Φ tal que X_k es $\Phi(k)$-conexo, f_k es una $2\Phi(k)$ equivalencia y $\Phi(k) - \frac{k}{2} \longrightarrow \infty$ si $k \longrightarrow \infty$. Una aplicación $g: X \longrightarrow Y$ de grado j es una familia de aplicaciones $g_k : X_k \longrightarrow Y_{k+j}$ tal que $g_k f_{k-1} = f_{k+j-1} \Sigma g_{k-1}$. Si no se hallara nada todas las aplicaciones son de grado 0.

$$\underline{F} \longrightarrow \underline{E} \longrightarrow \underline{B}$$

es una fibración de espectros si para cada k

$$F_k \longrightarrow E_k \longrightarrow B_k$$

es una fibración y los diagramas

$$\begin{array}{ccccc} F_k & \longrightarrow & E_k & \longrightarrow & B_k \\ \uparrow & & \uparrow & & \uparrow \\ \Sigma F_{k-1} & \longrightarrow & \Sigma E_{k-1} & \longrightarrow & \Sigma B_{k-1} \end{array}$$

son homotópicamente conmutativos.

Para una fibración de espectros obtenemos una sucesión exacta de cohomología

$$\longrightarrow H^q(\underline{B};G) \longrightarrow H^q(\underline{E};G) \longrightarrow H^q(\underline{F};G) \longrightarrow H^{q-1}(\underline{B};G).$$

Ejemplos de espectros

1) $\underline{S} = \{S^k\}$ espectro esférico.

2) $\underline{K(\pi,n)} = \{K(\pi,n+q)\}_{q\geq 0}$.

3) Si $\underline{E}$ es un espectro y X un espacio definamos un espectro $\underline{E}\wedge X$ mediante:

$$(\underline{E}\wedge X)_k = E_k \wedge X,$$

en particular:

3 a) $\underline{S}X = \underline{S}\wedge X$, la suspensión de X.

4) $\Omega\underline{X} = \{\Omega X_k\}$; $\mathcal{P}\underline{X} = \{\mathcal{P}X_k\}$.

5) $\Omega\underline{S}X \to \mathcal{P}\underline{S}X \to \underline{S}X$ es una fibración de espectros.

6) Si $\underline{F} \to \underline{E} \to \underline{B}$ es una fibración de espectros, y $g:\underline{X} \to \underline{B}$ se define la fibración de espectros inducida por g:

$$g^*(\underline{E}) = \{g_k^*(E_k)\}$$

6 a) Si $\alpha\beta = 0$ es una relación de grado m+1 entre operaciones primarias, $\beta \in A_n$, β está representada por $\beta:\underline{K(\underline{Z}_2,0)} \to \underline{K(\underline{Z}_2,n)}$ y tenemos una fibración inducida de la fibración de caminos

$$\begin{array}{ccc}
\underline{\Omega K(Z_2,n)} & & \underline{\Omega K(\underline{Z}_2,n)} \\
\downarrow i & & \downarrow \\
\underline{E} & & \underline{\mathcal{P}K(\underline{Z}_2,n)} \\
\downarrow & & \downarrow \\
\underline{K(Z_2,0)} & \longrightarrow & \underline{K(\underline{Z}_2,n)}
\end{array}$$

y podemos construir $u \in H^m(E)$ (coeficientes $\underline{Z}_2$ sobreentendidos donde $\bar{\gamma}_n$ es el generador de $H^*(\Omega K(Z_2,n))$ y está u determina en forma única una operación secundaria estable asociada con $\alpha\beta = 0$.

Una fibración de espectros es principal si para cada k, la correspondiente fibración de espacios es principal, y los diagramas que resultan son homotópicamente conmutativos.

Sea $\underline{X}$ un espectro, sean $n_o < n_1 < \ldots$ enteros tales que $\pi_{n_j}(\underline{X})$ son los grupos de homotopía no nulos de $\underline{X}$ y pongamos $\underline{K}_j = K(\pi_{n_j}(\underline{X}), n_{j+1})$.

Una torre de Postnikov para un espectro X es una torre de fibraciones principales

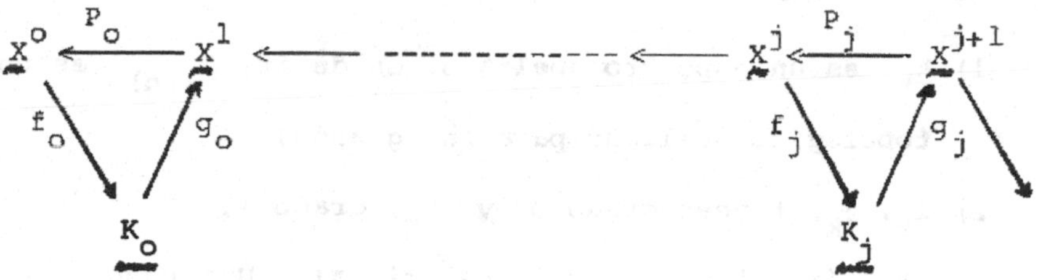

con f_k y g_k de grado 0 y p_k de grado 1, y tal que si $n_o < n_1 < n_2 < \ldots < n_t < \ldots$ son las dimensiones donde X tiene homotopía no trivial, entonces

$$\underline{K}_k = K(\pi_{n_k}(X), n_k+1)$$

y $\underline{X}^{\infty} \cong \text{lim proy } \underline{X}^j = \underline{X}$.

Mas generalmente, si $\underline{F} \to \underline{E} \to \underline{B}$ es una fibración de espectros, una torre de Postnikov para la fibración, es una torre de fibraciones principales.

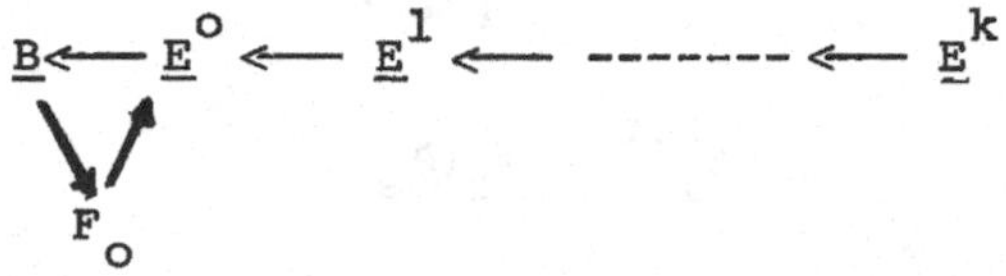

con $\underline{E}^{\infty} = \text{lim proy } E^j = \underline{E}$, y tal que si se construye la torre inducida por la inclusión de un punto en $\underline{B}$, resulta una torre de Postnikov para $\underline{F}$.

Sea $\underline{X}$ un espectro. Una <u>resolución geométrica</u> de $\underline{X}$ es una torre de fibraciones principales

$$
\begin{array}{ccccccccc}
\underset{\sim}{X} & \xleftarrow{P_o} & \underset{\sim}{X}_o & \longleftarrow & \underset{\sim}{X}_1 & \longleftarrow \cdots & \underset{\sim}{X}_k & \xleftarrow{P_{k+1}} & X_{k+1} \longleftarrow \cdots \\
 & {\scriptstyle i_o}\searrow & \nearrow{\scriptstyle j_o} & & & & {\scriptstyle i_k}\searrow & \nearrow{\scriptstyle j_k} & \\
 & & \underline{A}_o & & & & & \underline{A}_k &
\end{array}
\qquad (3.1)
$$

tal que:

1) $\underline{A}_k$ es un espectro abeliano, es decir, $\underline{A}_{k(n)}$ es un grupo topológico abeliano para n grande,

2) i_k, j_k, tienen grado 0 y P_k, grado 1,

3) $P_k^*:H^*(\underline{X}_{k-1}) \longrightarrow H^*(\underline{X}_k)$ es trivial. Una resolución geométrica se llama de <u>Adams</u> si además:

4) A_k es un producto de espectros $K(\underline{Z}_2,n)$.

Tomando grupos de homotopía en (3.1), obtenemos una cupla exacta [7]

$$
\begin{array}{ccc}
D_1 & \xrightarrow{P_*} & D_1 \\
{\scriptstyle j_*}\nwarrow & & \swarrow{\scriptstyle i_*} \\
 & E_1 &
\end{array}
$$

donde

$$D_1^{s,t} = \pi_t(\underset{\sim}{X}_s)$$

$$E_1^{s,t} = \pi_t(\underset{\sim}{A}_s).$$

Si la resolución geométrica es de Adams

$$E_1^{s,t} = \mathrm{Hom}_A^t(H^*(\underset{\sim}{A}_s),\underline{Z}_2)$$

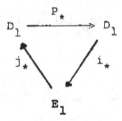

y la sucesión espectral deducida de esta cupla se llama la <u>sucesión espectral de Adams</u>. Es fácil ver que $E_2^{s,t} = \mathrm{Ext}_A^{s,t}(H^*(X),\underline{Z}_2)$. Esta sucesión espectral da información sobre la 2-componente de la homotopía del espectro $\underline{X}$ a partir de la estructura de A-módulo de $H^*(\underline{X})$. Para sus propiedades remitimos a [1].

Una <u>torre de Postnikov modificada</u> es un diagrama de fibraciones principales

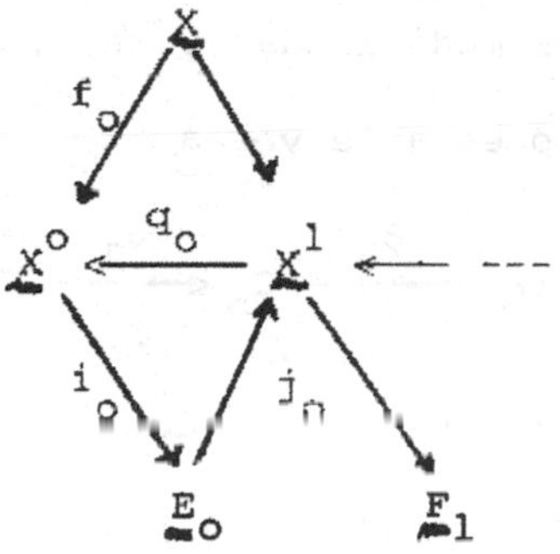

tal que:

1) $\underline{F}_k$ es abeliano,

2) f_k, i_k, j_k, tienen grado 0, q_k, grado 1, y $f_k = q_{k+1} f_{k+1}$,

3) $f_o^*: H^*(\underline{X}^o) \longrightarrow H^o(\underline{X})$ es epimorfismo,

4) $\text{Ker } f_k^* = \text{Ker } q_{k+1}^*$,

5) si además: $\underline{F}_k$ es un producto de $K(\underline{Z}_2, n)$ se llama una <u>torre modificada de Mahowald</u>,

6) si además f_k induce isomorfismo en homotopía para k grande se llama <u>convergente</u>.

Si $\underline{X}$ es un espectro y

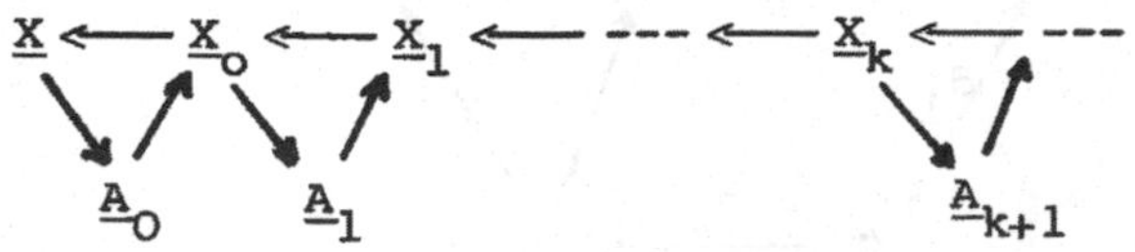

es una resolución geométrica, al tomar la inclusión de un punto x_o e inducir la torre sobre él, obtenemos una torre modificada para $\Omega\underline{X}$, por lo que tenemos así:

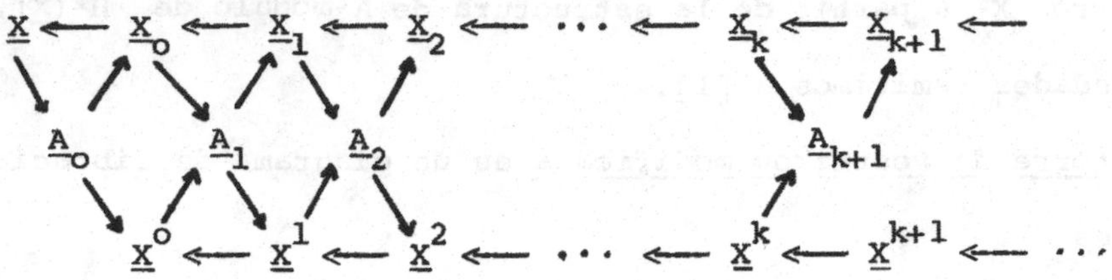

la relación entre torres modificadas y resoluciones geométricas.

Sea $\underline{X}$ un espectro estable y sea

$$0 \longleftarrow H^*(\underline{X}) \xleftarrow{\partial_o} C_o \xleftarrow{\partial_1} C_1 \xleftarrow{\partial_2} \cdots$$

una resolución $\mathcal{G}$-libre de $H^*(\underline{X})$, es decir, una sucesión exacta tal que cada C_i es un $\mathcal{G}$-módulo libre; a partir de ella construiremos una resolución geométrica de Adams para $\underline{X}$.

Sea $V_i = C_i \otimes_{\mathcal{G}} \underline{Z}_2$ y definamos $\underline{A}_i = K(V_i)$ que es un espectro abeliano.

Ahora bien, $H^q(\underline{X})$ está representado por las clases de homotopía de aplicaciones de $\underline{X}$ en el $K(\underline{Z}_2,q)$, luego existe una aplicación $f_o : \underline{X} \longrightarrow \underline{A}_o$ tal que $f_o^* = \partial_o$. Llamemos $\underline{X}_o$ a la fibra, que también es un espectro estable.

Pasando a cohomología tenemos la siguiente situación:

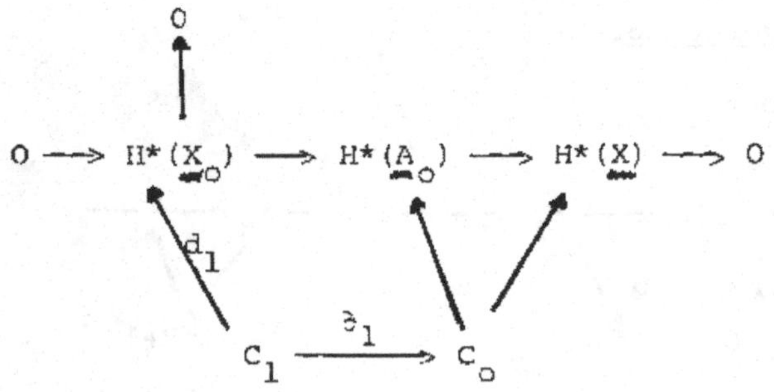

donde d_1 está determinado por la exactitud de las sucesiones, entonces aplicamos a $d_1 : C_1 \to H^*(\underline{X}_o)$ el procedimiento anterior, existe f_1 tal que $f_1^* = \partial_1$, llamemos $\underline{X}_1$ a la fibra, etc.

<u>Teorema</u> 3.2. <u>Sea $\underline{F} \xrightarrow{i} \underline{E} \xrightarrow{p} \underline{B}$ una fibración estable, entonces admite torres modificadas.</u>

<u>Demostración</u>. Tenemos la sucesión exacta de Serre,

$$\longrightarrow H^q(\underline{F}) \longrightarrow H^{q+1}(\underline{B}) \longrightarrow H^{q+1}(\underline{E}) \longrightarrow H^{q+1}(\underline{F}) \longrightarrow \cdots$$

Sea $Q = \mathrm{Im}(H^*(\underline{E}) \longrightarrow H^*(\underline{F}))$ y $S = \mathrm{Im}(\tau : H^*(\underline{F}) \longrightarrow H^*(\underline{B}))$ y sea Q_o un espacio vectorial graduado generado por un sistema de generadores de Q sobre A y Q_o^+, isomorfo a Q_o pero con un isomorfismo de grado -1. Sea S_o, definido analogamente a Q_o. Definimos

$$\underline{B} \xrightarrow{f_o} K(Q_o^+) \times K(S_o) = \underline{A}_o$$

tal que f_o^* es trivial en Q_o^+ y es monomorfismo en S_o. Sea $\underline{E}^o \longrightarrow \underline{B}$ el espacio fibrado inducido por f. Entonces $f_o p$ es trivial y p se levanta a $q_o : \underline{E} \longrightarrow \underline{E}^o$. Sea $\underline{F}_o$ la fibra de q_o. Tenemos el diagrama:

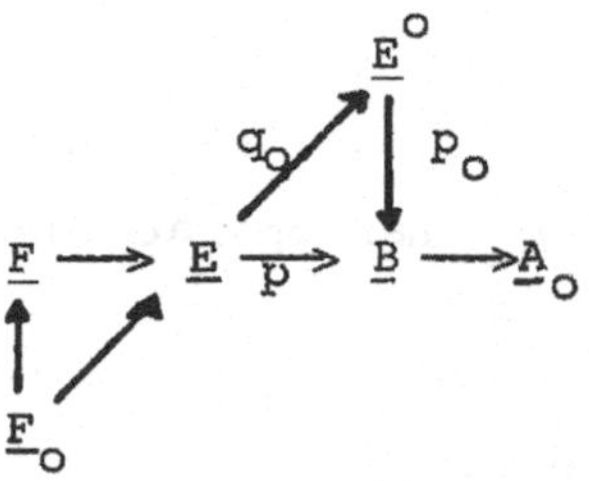

Podemos escoger q_o tal que q_o^* es epimorfismo. Entonces la transgresión en $\underline{F}_o \to \underline{E} \to \underline{E}^o$ es inyectiva. Construimos una aplicación de $\underline{F}_o$ en un producto de $K(Z_2, n)$ que sea epimorfismo en cohomología, y sea A_1 el espacio clasificante de este producto de $K(Z_2, n)$. Construimos $\underline{E}^o \xrightarrow{f_1} \underline{A}_1$ que en cohomología cubre a la transgresión. Entonces $f_1 p_o$ es nulhomotópico y p_o levanta a $\underline{E} \to \underline{E}^1$ donde $\underline{F}^1$ es el fibrado inducido por f_1. Procedemos así sucesivamente:

Si consideramos una fibración:

$$\underline{F} \to \underline{E} \xrightarrow{p} \underline{B}$$

tal que p^* es epimorfismo, entonces podemos construir una torre modificada a partir de una resolución algebraica de $H^*(F)$,

$$0 \leftarrow H^*(F) \leftarrow C_o \leftarrow C_1 \leftarrow C_2 \leftarrow \cdots$$

poniendo $\underline{A}_k = K(C_k \otimes_A Z_2)$ y construyendo transformaciones $f_k : \underline{E}^{k-1} \to A_k$ como arriba.

4. Operaciones de orden superior

Sea $C : C_o \xleftarrow{\partial_1} C_1 \xleftarrow{\partial_2} --- \xleftarrow{\partial_N} C_N$ un complejo libre sobre G, con las ∂_i de grado 0.

Una pirámide de operaciones cohomológicas $\Phi^{r,s}$, $n \geq r > s \geq 0$ asociada con C es una familia de operaciones que satisface los siguientes axiomas.

Axioma 1 (Inducción). Las operaciones $\Phi^{r,s}$ para $u \geq r \geq s \geq v$ están asociadas con

$$C_v \longleftarrow --- \longleftarrow C_u$$

Axioma 2 (Dominio). Si $\varepsilon : C_o \to H^*(\underset{\sim}{X})$ es un G-morfismo y $\Phi^{i,0}_{(\varepsilon)}$ está definida y contiene al cero para $i < N$, entonces $\Phi^{N,0}_{(\varepsilon)}$ está definida.

Axioma 3 (Valores). Si $\varepsilon : C_o \longrightarrow H^*(\underline{X})$ es un G-morfismo de grado y $\Phi^{N,O}(\varepsilon)$ está definida, entonces $\Phi^{N,O}(\varepsilon)$ es una clase de equiva-encia de G-morfismos $\eta : C_N \longrightarrow H^*(\underline{X})$ de grado $q-N+1$, para la elación de equivalencia $\eta \sim \eta'$ si $\mu \in \Phi^{N,1}(\rho)$ tal que $\eta^1 = \eta + \mu$ ara algún $\rho : C_1 \longrightarrow H^*(\underline{X})$ tal que $\Phi^{N,1}(\rho)$ está definido.

Axioma 4 (Naturalidad). Si $f : \underline{Y} \longrightarrow \underline{X}$ es una aplicación y $^{N,O}(\varepsilon)$ está definida para un $\varepsilon : C_o \longrightarrow H^*(X)$, entonces $\Phi^{N,O}(f^*\varepsilon)$ stá definida y

$$f^* \Phi^{N,O}(\varepsilon) = \Phi^{N,O}(f^*(\varepsilon)).$$

Axioma 5 (Peterson-Stein). Si $(\underline{X},\underline{Y})$ es un par, $\varepsilon : C_o \longrightarrow H^*(X)$ s tal que $\Phi^{N-1,O}(\varepsilon)$ está definida y $\exists\ \mu : C_{n-1} \longrightarrow H^*(\underline{X},\underline{Y})$ tal que $^*\mu \in \Phi^{N-1,O}(\varepsilon)$ entonces $\Phi^{N,O}(i^*\varepsilon)$ está definido; y para toda tal existe $\eta \in \Phi^{N,O}(i^*\varepsilon)$ tal que $\delta\eta = \Phi^{N,N-1}(\mu)$.

Una realización del complejo C es una torre de fibraciones rincipales:

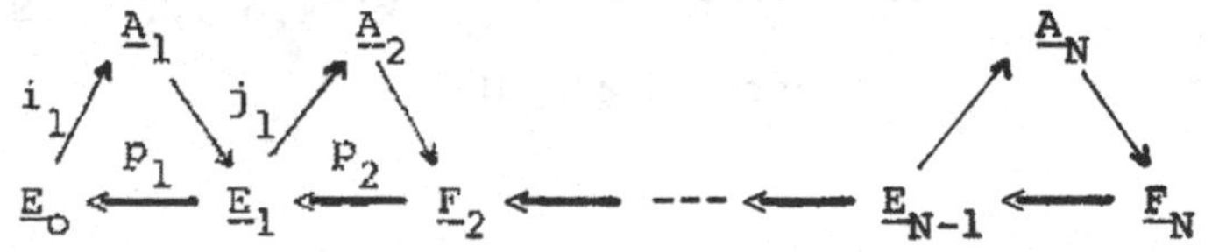

onde i_k, j_k tienen grado 0 y p_k tiene grado +1, con somorfismos:

$$\alpha : C_o \longrightarrow H^*(E_o) \quad \text{y si} \quad k > 0,$$

$$\alpha_k : C_k \longrightarrow H^*(\underline{A}_k)$$

ales que

$$j^*_{k-1} i^*_k \alpha_k = \alpha_{k-1} \partial_k$$

y

$$i^*_k \alpha_k \partial_{k+1} = 0.$$

Dado un complejo C, puede no tener una piramide asociada, pero si C tiene una realización, entonces un ejemplo universal para $\Phi^{N,0}$ es $i^*_N \alpha_N : C_N \longrightarrow H^*(E_{N-1})$. En general, dadas dos realizaciones geométricas las operaciones asociadas $\Phi^{N,0}$ no están relacionadas entre sí.

Definimos a continuación un complejo que veremos admite realizaciones geométricas. Las operaciones que se obtienen, generalizan a las operaciones secundarias Φ_{2r}.

Sea $C(N,r)$ el complejo:

$$C_o \xleftarrow{\partial_1} C_1 \longleftarrow \cdots \xleftarrow{\partial_N} C_N$$

donde

C_o tiene A-base c_o

C_N tiene A-base $c_{N,o}$

C_k tiene A-base $c_k, c_{k,o}, \ldots, c_{k,N-k}$

para $i \leq k \leq N-1$

con fronteras:

$$\partial_n \bar{c}_n = Sq^1 c_{n-1}$$
$$\partial_n c_{n,k} = Sq^1 c_{n-1,k} + Sq^{01} c_{n-1,k+1} + Sq^{2r-2k} c_{n-1}$$

donde $c_{o,k} = 0$ si $k \geq 0$. Si designamos por Φ^N_{2r} la operación asociada con $C(N,2r)$, entonces la relación que produce a Φ^{N+1}_{2r} es:

$$Sq^1 \Phi^N_{2r} + Sq^{01} \Phi^N_{2r-2} + Sq^{2r} \Phi^N_1 = 0$$

donde Φ^N_1 es la operación de orden n determinada por el Bockstein;

inductivamente Φ_1^N está asociada con la relación $Sq^1 \Phi_1^{N-1} = 0$.

Para construir una realización geométrica de $C(N,r)$ consideremos el espectro bu, donde

$$bu_{2q} = BU[2q]$$
$$bu_{2q+1} = U[2q+1]$$

donde $X[q]$ es el espacio que se obtiene de X al matar los primeros $q-1$ grupos de homotopía.

Sea $X_{2n+1} = bu \wedge \Sigma P^{\infty}_{2n+1}$ y E el algebra exterior generada por Sq^1 y $Sq^{01} = [Sq^1, Sq^2]$. Entonces $\mathrm{Ext}_E(Z_2, H^*(\Sigma P^{\infty}_{2n-1})$ tiene la siguiente forma:

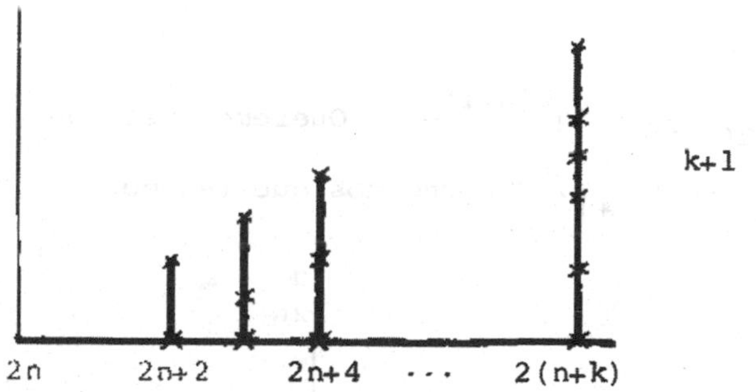

De aquí, utilizando el teorema sobre cambios de anillos, resulta que una resolución algebraica para $H^*(\underline{X}_{2n+1})$,

$$0 \longleftarrow H^*(\underline{X}_{2n-1}) \overset{\varepsilon}{\longleftarrow} C_0 \longleftarrow C_1 \longleftarrow --- \longleftarrow C_k \longleftarrow ---$$

es tal que C_k tiene generadores $c_{k,i}$, $i \geq 0$ en dimensiones $2(n+k)+k+2i$ y $\partial : C_k \longrightarrow C_{k-1}$ está dado por:

$$\partial c_{k,i} = Sq^1 c_{k-1,i} + Sq^{01} c_{k-1,i-1}.$$

Por (3.2), existe una torre de Mahowald para $\underline{X}_{2n+1}$,

$$\begin{array}{ccccccccccc}
\underline{A}_o = \underline{X}^o_{2n+1} & \longleftarrow & --- & \longleftarrow & \underline{X}^k_{2n+1} & \longleftarrow & \underline{X}^{k+1}_{2n+1} & \longleftarrow & --- & \longleftarrow & \underline{X}^{\infty}_{2n+1} = \underline{X}_{2n+1} \\
& \searrow & & \nearrow & & \searrow \quad \nearrow & & \searrow & & \nearrow & \\
& & \underline{A}_1 & & & \underline{A}_{k+1} & & & \underline{A}_{k+2} & &
\end{array}$$

Los k-invariantes en $\underline{X}^{k-1}_{2n+1}$ son clases $v_{k,i}$, $i \geq 0$ de dimensión $2(n+k+i)+1$ tales que

(4.1) $$Sq^1 v_{k,i} + Sq^{01} v_{k,i-1} = 0$$

En el caso de $\underline{X}^o_{2n+1}$, que es un producto de $K(Z_2, 2n+2i)$, $v_{o,i} = Sq^1 \alpha_{2(n+i)} + Sq^{01} \alpha_{2(n+i-1)}$, donde $\alpha_{2(n+i)}$ es el generador de $H^*(K(Z_2, 2n+2i))$. Consideremos la transformación:

$$K(Z,O) \xrightarrow{f_o} \underline{A}_o$$

donde $f^*_o \alpha_{2(n+i)} = Sq^{2(n+i)} \gamma_o$. Queremos ver que f_o levanta a $f_\infty : K(Z,O) \longrightarrow \underline{X}_{2n+1}$. Supongamos que tenemos

$$\begin{array}{ccccc}
& & \underline{X}^q_{2n+1} & & \\
& & \downarrow & & \\
K(Z,O) & \xrightarrow{f_{q-1}} & \underline{X}^{q-1}_{2n+1} & \longrightarrow & \underline{A}_q \\
& \searrow^{f_{q-2}} & \downarrow & & \\
& & \underline{X}^{q-2}_{2n+1} & \longrightarrow & \underline{A}_{q-1}
\end{array}$$

entonces

$$f^*_{q-1} v_{q,i} = \alpha_{q,i}, \ \alpha_{q,i} \in A/ASq^1$$

y de (4.1), obtenemos $Sq^1 \alpha_{q,i} + Sq^{0,1} \alpha_{q,i-1} = 0$, un sistema de ecuaciones. En [5], hemos visto que en este caso:

$$\left.\begin{array}{l}
\alpha_{q,i} \equiv Sq^1 \alpha^1_{q,i} + Sq^{01} \alpha^2_{q,i} \\
\alpha^2_{q,i} = \alpha^1_{q,i-1}
\end{array}\right\} \quad \text{mod } A\, Sq^1$$

Luego los valores de $f^*_{q-1}v_{q,i}$ pueden alterarse, variando f_{q-2}, tal que simultaneamente $f^*_{q-1}v_{q,i} = 0$, y podemos extender a $\underline{X}^q_{2n+1}$. El ejemplo mas universal de orden q se construye tomando el espacio fibrado inducido por f_q del espacio de curvas de X^q_{2n+1}.

En [5], estudiamos los valores en dimensiones bajas de cualquier operación Φ^N_{2r} asociada con el complejo $C(N,2r)$ y en particular obtenemos: Si $x \in H^q(X)$ proviene de una clase entera y $q \leq 2r-4N+4$, entonces $\Phi^N_{2r}(X)$ está definida y es una operación primaria estable en x.

D. Hurley, en su tesis ha generalizado el teorema (2.4) correspondiente a evaluación en términos de clases de Chern.

5. Dualidad

Sean X, Y CW-complejos, una N-dualidad entre ellos es una transformación:

$$\varphi : X \wedge Y \longrightarrow S^N \tag{5.1}$$

tal que $\varphi^* i_N / : H_q(X) \longrightarrow H^{N-q}(Y)$ es isomorfismo para toda q. Decimos que φ es N S-dualidad si se tiene el isomorfismo al suspender a X y Y convenientemente. Si X es subcomplejo de S^{N+1} y X^* es S-retracto por deformación de $S^{N+1}-X$, entonces X y X^* son N S-duales. Luego todo complejo finito tiene S-duales que tienen el mismo S-tipo. Si X es complejo, designemos por X^* un S-dual de X, entonces si X y X^* son S-duales, $X \wedge X^*$ es auto-dual. Tenemos $\{X,Y\} \cong \{Y^*,X^*\}$. Si X y Y son N S-duales, podemos obtener una S-aplicación

$$S^N \xrightarrow{\psi} X \wedge Y$$

tal que para cada primo p,

$$H^r(X) \otimes H^{N-r}(Y) \xrightarrow{\psi^*} H^N(S^N)$$

es un apareamiento no singular.

Consideremos $p = 2$ y designemos el apareamiento por $\langle\ ,\ \rangle$. Si $\theta \in A_q$, definimos $x(\theta) \in A_q$ por

$$\langle x(\theta)x,y\rangle = \langle x,\theta(y)\rangle$$

para toda $x \in H^r(X)$, $y \in H^{N-r-q}(X^*)$.

Lema 5.1. La transformación $x:A \longrightarrow A$ es un antiautomorfismo que cumple con $X(Sq)Sq = 1$, donde

$$Sq = \sum_{i=o}^{\infty} Sq^i$$

$$X(Sq) = \sum_{i=o}^{\infty} X(Sq^i).$$

Demostración.

1) $\langle x(\theta_1\theta_2)x,y\rangle = \langle x,\theta_1\theta_2 y\rangle = \langle x(\theta_1)x,\theta_2 y\rangle$
$= \langle x(\theta_2)x(\theta_1)x,y\rangle$

2) Sean $x \otimes y \in H^{N-1}(X \wedge X^*)$, entonces

$$\psi^*(Sq^i(x \otimes y)) = 0 = \sum_{j=o}^{i} \langle Sq^j x, Sq^{i-j} y\rangle$$

$$= \sum_{j=o}^{i} \langle X(Sq^{i-j})Sq^j x,y\rangle$$

y el resultado se sigue.□

Queremos proceder ahora con operaciones de orden superior. Sea $\alpha\beta = 0$ una relación y Φ una operación secundaria asociada con esta relación.

Queremos definir $X(\Phi)$ por:

$$\langle X(\Phi)x,y\rangle = \langle x,\Phi(y)\rangle$$

uego $y \in N(\beta)$, además $\Phi(y) \ni 0$ si $\alpha z \in \Phi(y)$, luego $x(\alpha)x = 0$. demás los valores $x(\Phi)(x)$ varian por $\mathrm{Im}(X(\beta))$, luego $X(\Phi)$ está sociada con $(\beta)(\alpha) = 0$.

Queremos generalizar esta dualidad a operaciones de orden N. rimero veamos el complejo dual: Si

$$C: C_o \longleftarrow C_1 \longleftarrow --- \longleftarrow C_k \longleftarrow --- \longleftarrow C_N$$

s un complejo, construimos su complejo dual como sigue:

Si ponemos un functor de A-módulos en A-modulos que a todo A-ódulo M asocia el A-módulo $\bar{M} = \mathrm{Hom}_A(M,A)$ con estructura de A-ódulo

$$(a\cdot f)(m) = f(m)\chi(a)$$

a todo $f: M \longrightarrow N$ asocia $f^*: \bar{N} \longrightarrow \bar{M}$ dado por $f^*(f)(m) = f(\varphi(m))$. i M es A-módulo libre con base $\{m_i\}$, decimos que $\bar{m}_j$ es base ual de $\bar{M}$ si $\bar{m}_j(m_i) = \delta_{ji}$. Es fácil ver

<u>Lema</u> 5.2. <u>Si</u> M <u>es libre con base</u> $\{m_i\}$, $\bar{M}$ <u>es libre con base</u> $\{\bar{m}_i\}$. <u>Si</u> M_1, M_2 <u>son libres con bases</u> $\{m_i\}$, $\{n_j\}$ <u>y</u> $f: M_1 \longrightarrow M_2$ <u>stá dado por</u>

$$f(m_i) = \Sigma a_{ij} n_j$$

<u>ntonces</u>

$$\bar{f}(n_j) = \Sigma\,(a_{ij})\bar{m}_i$$

efinimos $\bar{C}: \bar{C}_o \xrightarrow{\partial} \bar{C}_1 \longrightarrow --- \xrightarrow{\bar{\partial}_N} \bar{C}_N$ como el complejo dual a C.

Si $\Phi^{r,s}$ es una piramide de operaciones cohomológicas asociada con C, $(\bar{\Phi}^{r,s})$ da una piramide de operaciones cohomológicas asociada con $\bar{C}$.

Ejercicio 5.3. Describir explícitamente el complejo $\bar{C}(N,2r)$, dual de $C(2r,N)$.

Sea $\mathcal{S}$ = la categoría de espectros estables $\underline{Y}$ tal que $\Pi_i(Y)$ es finito para toda i y clases de homotopía. Si X es un CW-complejo $\{X_\alpha\}$ denota la familia de subcomplejos finitos de X, $i_\alpha : X_\alpha \to X$ las inducciones. Recordemos que para $\underline{h} \in \mathcal{S}$, definimos grupos de homología y cohomología de X, como

$$\underline{h}_q(X) = \varinjlim_n \Pi_{q+n}(X \quad h_n)$$

y

$$\underline{h}^q(X) = \varprojlim_n [\Sigma^{q+n}X, h_n]$$

Teorema 5.4. Si $h \in \mathcal{S}$, tenemos

$$\varinjlim h_q(i_\alpha) : \varinjlim h_q(X_\alpha) \cong h_q(X)$$

$$\varprojlim h^q(i_\alpha) : h^q(X) \cong \varprojlim h^q(X_\alpha).$$

Tenemos que $h^q(X_\alpha)$ es finito, luego compacto y damos a $h^q(X)$ la topología compacta inducida por $\varprojlim h^q(X_\alpha)$. Sea G un grupo abeliano, designamos por $G^t = \mathrm{Hom}(G,S^1)$ su grupo de caracteres.

Si $h \in \mathcal{S}$, sea $(h^t)^q$ el functor de CW-complejos en grupos compactos dado por:

$$(h^t)^q(X) = h_q(X)^t = \varprojlim h_q(X_\alpha)^t$$

entonces h^t define una teoría de cohomología aditiva, luego por el teorema de Brown [11] existe $\chi(h) \in \mathcal{S}$ tal que

$$T_h\colon (h)^q(X) \cong (h^t)^q(X) \cong (h_q(X))^t$$

Si $h,k \in \mathcal{S}$ y $u \in [h,k]_q$, existe un único $\chi(\mu) \in [\chi(k),\chi(h)]_q$ tal que:

$$\begin{array}{ccc} \chi(k)^q(X) & \xrightarrow{T_k} & k_q(X)^t \\ \Big\downarrow (\mu)_* & & \Big\downarrow (\mu_*)^t \\ \chi(h)^q(X) & \xrightarrow{T_u} & h_q(X)^t \end{array}$$

conmuta.

<u>Teorema</u> 5.5. χ <u>tiene las siguientes propiedades</u>:

i) χ <u>es functor contravariante</u> $\mathcal{S} \longrightarrow \mathcal{S}$.

ii) <u>Para cada</u> $h \in \mathcal{S}$, <u>existe una equivalencia natural</u> $S_h\colon \chi(h)^t_q \longrightarrow h^q$ <u>y que es natural respecto a</u> h.

iii) $\chi\chi$ <u>es naturalmente equivalente con</u> $1\colon \mathcal{S} \longrightarrow \mathcal{S}$.

iv) $\chi\colon [h,h]_q \cong [\chi(k),\chi(h)]_q$.

v) $\chi\colon [K(Z_2,0),K(Z_2,0)] \longrightarrow [K(Z_2,0),K(Z_2,0)]$ <u>es el antiautomorfismo canónico de</u> A

<u>Demostración</u>. Para definir

$$S_h\colon \chi(h)^t_q \longrightarrow h^q$$

basta suponer X CW-complejo finito y sea X^* un m S-dual de X. Tenemos

$$\chi(h)_q(X)^t \cong \chi(h)^{m-q}(X^*)^t \cong h_{m-q}(X^*) \cong h^q(X)$$

y el resultado se sigue por el trabajo de G.W. Whitehead en dualidad de Alexander. iii) es consecuencia de que tenemos $S_h T_{\chi(h)}$ da una equivalencia entre $\chi\chi(h)^q$ y h^q. Veamos v). Supongamos $a \in A = [K(Z_2,0),K(Z_2,0)]_i$. Si X^* es un m S-dual de X,

$$\begin{array}{ccccc}
H^{m-q}(X^*)^t & \approx & H_q(X)^t & \cong & H^q(X) \\
\Big\uparrow {\scriptstyle a^t} & & \Big\uparrow {\scriptstyle \bar{a}^t} & & \Big\downarrow {\scriptstyle u} \\
H^{m-q+i}(X^*)^t & \cong & H_{q-i}(X)^t & \cong & H^{q+i}(X)
\end{array}$$

y es bien sabido que $u = \chi(a)$, (ver el Lema (5.1)).

Sea

$$\begin{array}{cccccccccccc}
F_{k,k-1} & \rightarrow & F_{k,k-2} & \rightarrow \cdots \rightarrow & F_{k,1} & \rightarrow & F_{k,o} & \rightarrow & E_k & & \\
& & \downarrow & & \downarrow & & \downarrow & & \downarrow & & \\
& & F_{k-1,k-2} & \rightarrow \cdots \rightarrow & F_{k-1,1} & \rightarrow & F_{k-1,o} & \rightarrow & E_{k-1} & \rightarrow & BF_{k,k-1} \\
& & \downarrow & & \downarrow & & \downarrow & & \downarrow & & \\
& & F_{k-2,k-3} & \rightarrow \cdots \rightarrow & F_{k-2,1} & \rightarrow & F_{k-2,o} & \rightarrow & E_{k-2} & & \\
& & & & & & \downarrow & & \downarrow & & \\
& & & & & & \vdots & & \vdots & & \\
& & & & & & & & \downarrow & & \\
& & & & & & F_{1,o} & \rightarrow & E_1 & & \\
& & & & & & & & \downarrow & & \\
& & & & & & & & E_o & &
\end{array}$$

un ejemplo universal para una operación de orden k. Tomando χ obtenemos:

$$
(5.7)\qquad
\begin{array}{ccccccccccc}
\chi(E_o) & \to & \chi(E_1) & \to \cdots \to & \chi(E_{k-2}) & \to & \chi(E_{k-1}) & \to & \chi(E_k) & & \\
\downarrow & & & & & & \downarrow & & \downarrow & & \\
 & & \chi(F_{1,o}) & \to \cdots \to & \chi(E_{k-2,o}) & \to & \chi(F_{k-1,o}) & \to & \chi(F_{k,o}) & \to & B\chi(E_o) \\
 & & & & & & \downarrow & & \downarrow & & \\
 & & & & \chi(F_{2,1}) \to \cdots \to \chi(F_{k-2,1}) & \to & \chi(F_{k-2,1}) & \to & \chi(F_{k,1}) & & \\
 & & & & & & \downarrow & & \downarrow & & \\
 & & & & & & \vdots & & \vdots & & \\
 & & & & & & \downarrow & & \downarrow & & \\
 & & & & & & \chi(F_{k-1,k-2}) & \to & \chi(F_{k,k-2}) & & \\
 & & & & & & & & \downarrow & & \\
 & & & & & & & & \chi(F_{k,k-1}) & &
\end{array}
$$

que resulta también un ejemplo universal para una operación de orden k.

Observese que si E_{k-1} es el ejemplo universal para la operación asociada con (5.6), $\chi(F_{k.o})$ es el ejemplo universal para la operación asociada con (5.7). Es fácil ver que si (5.6) tiene por complejo asociado

$$C_o \xleftarrow{\partial_1} C_1 \xleftarrow{\partial_2} \cdots \xleftarrow{\partial_k} C_k$$

entonces (5.7) tiene por complejo asociado:

$$\chi(C_k) \xleftarrow{\chi(\partial_k)} \chi(C_{k-1}) \longleftarrow \cdots \longleftarrow \chi(C_1) \xleftarrow{\chi(\partial_1)} \chi(C_o)$$

Se puede entonces verificar que si ^{N,O} está asociada con (5.6) y $\psi^{N,O}$ con (5.7), entonces para CW-complejos duales X, Y y clases x, y en $H^*(X)$, $H^*(Y)$ respectivamente, que

$$\langle \Phi^{N,O}(x), y\rangle = \langle x, \psi^{N,O}(y)\rangle.$$

En particular podemos construir operaciones duales a las operaciones

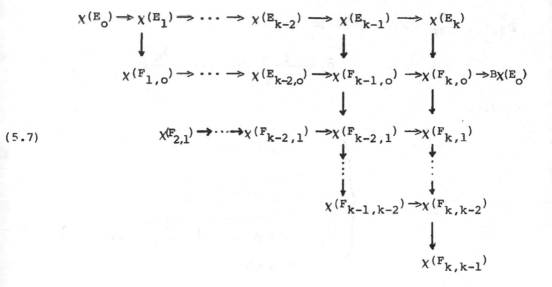

Φ^N_{2r}, $\chi(\Phi^N_{2r})$ y el complejo: $\bar{C}(N,2r)$

$$\bar{C}_o \xleftarrow{\bar{\partial}_1} \bar{C}_1 \longleftarrow \cdots \xleftarrow{\bar{\partial}_N} \bar{C}_N$$

donde

$$\begin{cases} \bar{C}_o \text{ tiene A-base } \bar{c}_o \\ \bar{C}_N \text{ tiene A-base } \bar{c}_N \\ \bar{C}_k \text{ tiene A-base } \{\bar{c}^N_k, \bar{c}_{k,o}, \ldots, \bar{c}_{k,k}\}, \\ \qquad 1 \leq k \leq N-1 \end{cases}$$

y los operadores $\bar{\partial}_i$ están dados por:

$$\begin{cases} \bar{\partial}_1 \bar{c}_1 = \chi(Sq^{2r}) \bar{c}_o \\ \bar{\partial}_1 \bar{c}_{1,o} = Sq^1 \bar{c}_o \\ \bar{\partial}_1 \bar{c}_{1,1} = Sq^{01} \bar{c}_o \end{cases}$$

y para $2 \leq k \leq N$,

$$\begin{cases} \bar{\partial}_k \bar{c}_k = Sq^1 \bar{c}_{k-1} + \sum_{t=o}^{k-1} \chi(Sq^{2r-2t}) \bar{c}_{k-1,t} \\ \bar{\partial}_k \bar{c}_{k,o} = Sq^1 \bar{c}_{k-1,o} \\ \bar{\partial}_k \bar{c}_{k,i} = Sq^1 \bar{c}_{k-1,i} + Sq^{01} \bar{c}_{k-1,i-1} \\ \bar{\partial}_k \bar{c}_{k,k} = Sq^{01} \bar{c}_{k-1,k-1} \end{cases}$$

es admisible.

Maunder en [8], demuestra que el complejo $C^*(N,2r)$ es admisible al construir operaciones cohomológicas que se relacionan con el caracter de Chern.

Problemas. 1.- Decidir si las operaciones Φ^N_{2r} y las operaciones de Maunder son duales.

2.- Calcular las operaciones Φ^N_{2r} en los espacios CP^n y QP^n.

3.- Calcular fórmulas de producto para estas operaciones.

REFERENCIAS

[1] Adams, J.F., Stable homotopy theory, Lecture Notes in Mathematics 3, Springer-Verlag, Berlín, 1964.

[2] Barcus, W.D. y Meyer, J.P., The suspension of a loop space, Amer. J. Math. 80(1958).

[3] Gitler, S. y Milgram, R.J., Evaluating secondary operations on low dimensional clases, Conf. Algebraic Topology, University of Illinois, Chicago Circle 1968.

[4] Gitler, S., Mahowald, M. y Milgram, R.J., Secondary cohomology operations and complex vector bundles, Proc. of A. M. S., 22(1969).

[5] Gitler, S. y Milgram, R.J., Unstable divisibility of the Chern character, Bol. Soc. Mat. Mex. 16(1971).

[6] Hussemoller, D., Fiber bundles, McGraw-Hill, New York, 1966.

[7] MacLane, Homology, Springer-Verlag 1963.

[8] Maunder, C.R.F., Cohomology operations of the Nth kind, Proc. London Math. Soc. 13(1963).

[9] Milgram, R.J., The Bac construction and Abelian H-spaces, Ill. J. Math. 11(1967).

[10] Mosher, R.E. y Tangora, M.C., Cohomology operations and applications in homotopy theory, Harper and Row, New York, 1968.

[11] Peterson, F.P. y Stein, N., Secondary cohomology operations; two formulas, Amer. J. Math. 81(1959).

[12] Peterson, F.P. y Thomas, E., A note on non-stable cohomology operations, Bol. Soc. Mat. Mex. Segunda serie 3(1958).

[13] Serre, J.P., Homologie singuliere des espaces fibrés, Ann. Math.

(2), 54(1951).

[14] Spanier, E.H., Algebraic topology, McGraw-Hill, 1966.

[15] Steenrod, N.E. y Epstein, D.B.A., Cohomology operations, Ann. Math. Studies 50, Princeton University Press, 1962.

TWO PROBLEMS STUDIED BY HEINZ HOPF

Ioan M. James

It is hardly possible to study algebraic topology for long without coming to appreciate how many good ideas are derived from the work of Heinz Hopf. Since this Summer School is dedicated to his memory, I devote these lectures to a report of recent progress in connection with two out of the many interesting problems which were at one time or another studied by Hopf.

Possibly the problem of finding Euclidean models of projective spaces is more facinating than important. But like any good problem it has, over the years, acted as a great stimulus for the development and testing of new methods and techniques. There are some striking achievements to report, as well as much useful work, even though the basic questions remain to a large extent unanswered. For reasons of space I have not attempted to give a step-by-step review of the progress of the subject but rather to give a sketch of the present position including (I hope) the best specific results which have so far been obtained. As an introduction the reader would do well to consult the two most relevant papers of Hopf [34], [35] (these numbers refer to the bibliography for the first part of these lectures). A stimulating account of recent work on embeddings and immersions has been given by Gitler [22]. These notes are based on my survey article [39] on the same subject, but the material has been rearranged and by reducing the length of the introduction it has been possible to deal with other matters more thoroughly, and to repair some omissions. Many improvements are due to the helpful comments of Professors Adem, Feder, Gitler, Lam and Mahowald, in Mexico, and to Drs. Rees and Steer, in Oxford. As well as these I would also like to thank those who provided preprints of work in process of publication.

The second part of these lectures is concerned with certain problems discussed by Hopf in the Courant Festschrift. The basic questions are as follows. Consider a sphere-bundle, for example the bundle of unit tangent vectors to a Riemannian manifold. When does there exist a fibre-preserving map of this sphere-bundle into itself which transforms each vector into an orthogonal vector? When does there exist a fibre-preserving homotopy which deforms each vector into its own opposite? In these lectures I give an introduction to research on these questions, including a few results directly inspired by the original article of Heinz Hopf.

PART I: EUCLIDEAN MODELS OF PROJECTIVE SPACES

1. Introduction

Due to limitations of space I will, in these lectures, restrict my attention to manifolds etc. which are smooth, in the C^{∞} sense. Let X be a compact connected manifold. Let $R^q (q=1,2,\ldots,)$ denote euclidean q-space. We write $X \subset R^q$ (resp. $X \subseteq R^q$) if X can be <u>embedded</u> (resp. <u>immersed</u>) in R^q. The fundamental papers of Whitney [81], [82] show that $X \subset R^{2m}$ and $X \subseteq R^{2m-1}$, where $m = \dim X$. It is interesting, in particular cases, to try and determine the least integer q such that $X \subset R^q$, or such that $X \subseteq R^q$.

We describe a vector bundle E over X as a <u>normal bundle</u> of

X if the direct sum $E \oplus T(X)$ is trivial, where $T(X)$ denotes the tangent bundle of X. The (mod 2) Stiefel-Whitney classes of a normal bundle E are determined by those of $T(X)$ and are denoted by $\bar{w}_i(X)$ $(i = 1,2,\ldots,)$ where $\bar{w}_i(X) \in H^i(X)$. Of course $\bar{w}_i(X) = 0$ for $i > \dim E$. An immersion of X in R^{m+k}, where $m = \dim X$, determines a normal k-plane bundle E. If the immersion is an embedding then E has various special properties, for example $\bar{w}_k(X) = 0$. By the fundamental theorem of Hirsch [33], if there exists a normal k-plane bundle of X, with $k \geq 1$, then there exists an immersion of X in R^{m+k}. In particular if X is parallelizable (i.e. if $T(X)$ is trivial) then there exists an immersion of X in R^{m+1}. The corresponding theorem for embeddings, due (amongst others) to Browder [13], is considerably more complicated. Some indication of the extensive literature on these and related matters can be found in the articles of Gitler [22] and James [39].

The purpose of these lectures, however, is not to discuss the general theory but to report progress in the special case where X is one of the projective spaces. There are various special features in this case and it has attracted great attention from the outset. Two papers of Hopf [34], [35] are particularly concerned with this problem and contain a number of ideas which have proved particularly fruitful, especially in relation to the results described in 2 below.

Consider the algebra K, over the reals, where K = R, the real numbers, K = C, the complex numbers, or K = H, the quaternions. We regard the n-dimensional projective space $KP^n (n = 0,1,2,\ldots)$ as a dn-manifold, in the standard way, where $d = \dim_R K$. Also we regard

KP^{n-1} as a submanifold of KP^n, where $n \geq 1$, so that $KP^n - KP^{n-1}$ is a dn-cell. Note that an embedding or immersion of KP^n determines, by restriction, an embedding or immersion of KP^{n-1}. For the sake of simplicity, inferences of this type will generally be omitted when specific results are being discussed.

The mod 2 cohomology ring of KP^n is a truncated polynomial ring of height $n+1$, generated by an element $u \in H^d(KP^n)$. Except in case $K = R$, the corresponding statement is also true for integral cohomology. The normal Stiefel-Whitney classes are given (see [12]) by

$$\bar{w}_{di}(KP^n) = \binom{n+i}{i} u^i. \tag{1.1}$$

If n is a power of two then the binomial coefficient in this formula is non-zero, mod 2, for all $i \leq n$, and so we obtain

__Theorem__ 1.2. __Let__ n __be__ __a__ __power__ __of__ __two__. __Then__ KP^n __cannot__ __be__ __embedded__ __in__ R^{2dn-d} __or__ __immersed__ __in__ $R^{2dn-d-1}$.

Embeddings of projective spaces can be constructed by direct methods. For example suppose that we represent points of KP^n in the usual way by vectors

$$z = (z_0, \ldots, z_n) \quad (z_0, \ldots, z_n \in K)$$

of unit modulus, with scalar action on the left. Write $w_0 = \bar{z}_0 z_0$ and, for $i = 1, \ldots, 2n-1$, write

$$w_i = \Sigma \bar{z}_s z_t \ (s+t = i, 0 \leq s \leq t \leq n).$$

The vector $(w_0, \ldots, w_n)$ defines a point of euclidean $(2dn-d+1)$-space, since the first component is real, and is independent of the choice of representative z for a given point of KP^n. The mapping thus defined is an embedding and so we obtain

Theorem 1.3. *For all values of* n *there exists an embedding of* KP^n *in* $R^{2dn-d+1}$.

A modification of the same construction shows that the Cayley projective plane, which is a 16-manifold, can be embedded in R^{25}. This type of construction was first used by Hopf [34], in the real case, and extended by James [37] to the other cases. However the rational formulae we have given are due to Vranceanu [79]. By similar methods Hopf [34] and James [37] have proved

Theorem 1.4. *Let* n *be odd and let* $n > 1$. *Then* $RP^n \subset R^{2n-1}$ *and* $CP^n \subset R^{4n-3}$.

Given a sphere-bundle over a manifold it is to be expected that there is some relation between the immersion problem for the base and that for the total space. Sanderson [68] has studied this relationship in the case of a principal bundle and has obtained the following application. Recall that if $m \geq 1$ then RP^{2m+1} fibres over CP^m and RP^{4m+3} fibres over HP^m. In both cases the fibration is that of a principal differentiable sphere-bundle and we obtain

Theorem 1.5. *If* $CP^m \subseteq R^q$ *then* $RP^{2m+1} \subseteq R^{q+1}$. *If* $HP^m \subseteq R^q$ *then* $RP^{4m+3} \subseteq R^{q+3}$.

With the help of these and other considerations we arrive at

<u>Theorem</u> 1.6. <u>If</u> $n > 1$ <u>then</u> $RP^n \subseteq R^{2n-1}$. <u>Also</u> $RP^3 \subseteq R^4$. <u>Let</u> $n > 3$. <u>If</u> $n \equiv 1$ <u>mod</u> 2 <u>then</u> $RP^n \subseteq R^{2n-3}$. <u>If</u> $n \equiv 3$ <u>mod</u> 4 <u>then</u> $RP^n \subseteq R^{2n-6}$.

Whitney's theorem gives the first result. The second follows from Hirsch's theorem since RP^3 is parallelizable. The other results are obtained by applying (1.5) to the embeddings of (1.4).

2. Special features of the real case

The real case presents a number of special features which require separate discussion. First of all RP^n has 2-torsion, for $n \geq 2$; moreover RP^n is non-orientable when n is even. The Hopf line bundle H over RP^n satisfies the relation $H \otimes H \approx 1$. The tangent bundle $T(RP^n)$ satisfies the relation

$$T(RP^n) \oplus 1 \approx (n+1)H. \tag{2.1}$$

Hence if E is a normal k-plane bundle of RP^n then

$$E \otimes H \oplus (n+1) \approx (n+k+1)H. \tag{2.2}$$

Following Sanderson [67] we refer to $E \otimes H$ as a <u>twisted normal bundle</u> of RP^n. Sanderson [67] (see also James [38]) notes that

$$(n+k+1)H \approx 1 \oplus T|RP^n,$$

where $T = T(RP^{n+k})$, and so we obtain

<u>Theorem</u> 2.3. <u>There exists an immersion of</u> RP^n <u>in</u> R^{n+k} <u>if and only if there exists over</u> RP^n <u>a field of orthonormal</u> n-<u>frames of tangents to</u> RP^{n+k}.

In [35] Hopf considers a number of problems which are related to those we have been discussing, including the following. Let $t \geq m$ and $t \geq n$. By an <u>axial map</u>

$$g : RP^m \times RP^n \to RP^t$$

we mean a map such that

$$g(x,e) = x, \; g(e,y) = y \; (x \in RP^m, y \in RP^n),$$

where e denotes the basepoint in all cases. For example, suppose that we have a bilinear map

$$h : R^{m+1} \times R^{n+1} \to R^{t+1}$$

We describe h as <u>non-singular</u> if $h(x,y)$ is non-zero whenever $x \in R^{m+1}$ and $y \in R^{n+1}$ are non-zero. When this condition is satisfied we obtain from h a map

$$g : RP^m \times RP^n \to RP^t,$$

which is homotopic to an axial map. For example, take $t = m+n$. Then a non-singular bilinear map h is given by

$$h(x,y) = z = (z_o, \ldots, z_{m+n}),$$

where

$$z_k = \sum_{i+j=k} x_i y_j \quad (0 \leq k \leq m+n)$$

For suitable values of m and n it is possible to find non-singular bilinear maps with lower values of t than $m+n$. This is shown by Adem [2], [3], Behrend [11], Hopf [34], Lam [42], [43], [44], [45] and Milgram [61], amongst others.

One of the earliest applications of cohomology theory is the proof by Hopf [35] that axial maps do not exist, for certain values of m, n and t. In modern terminology we take cohomology with mod 2 coefficients and identify

$$H^*(RP^m \times RP^n) = H^*(RP^m) \otimes H^*(RP^n),$$

in the usual way, so that a map

$$g: RP^m \times RP^n \longrightarrow RP^t$$

induces a (ring) homomorphism

$$g^*: H^*(RP^t) \longrightarrow H^*(RP^m) \otimes H^*(RP^n).$$

If g is axial then $g^*u = u \otimes 1 + 1 \otimes u$, where u means the same as in §1, and hence

$$g^*(u^{t+1}) = (g^*u)^{t+1} = (u \otimes 1 + 1 \otimes u)^{t+1}$$

$$= \sum_{r+s=t+1} \binom{r+s}{r} u^r \otimes u^s$$

Since $u^{t+1} = 0$ this establishes

Theorem 2.4. <u>If there exists an axial map of</u> $RP^m \times RP^n$ <u>into</u> RP^t <u>then the binomial coefficient</u> $\binom{t+1}{r}$ <u>is even, for</u> $t+1-n \leq r \leq m$.

For example, when $t+2$ is a power of two it follows that no axial map exists unless $t \geq m+n$. Further necessary conditions for the existence of axial maps have been found by James [38], using the Adams ψ-operations of K-theory, and recently by Davis [15] using powerful techniques of the Postnikov type.

It was pointed out by James [38] that a general solution of the axial map problem would enable us to determine the geometric dimension (see [8]) of all vector bundles over real projective space, and there are strong implications from the latter problem back to the former. In particular Sanderson [67] has noted

Theorem 2.5. If there exists an immersion of RP^n in R^{n+k} then there exists an axial map of $RP^n \times RP^n$ in RP^{n+k}. Moreover the converse is true when $2k > n$.

In fact the converse of (2.5) is true without exception. The proof is somewhat indirect and will be given in a forthcoming publication. Although the result is less strong, a more direct proof can be given of

Theorem 2.6. If there exists a non-singular bilinear map $R^{n+1} \times R^{n+1} \to R^{n+k+1}$ then there exists an immersion of RP^n in R^{n+k}

This result of Ginsburg [20] is proved as follows. A bilinear map h, as above, determines a map

$$h' : S^n \times R^{n+1} \to S^n \times R^{n+k+1},$$

where the first component is the natural projection and the second is h. Clearly h induces a homomorphism

$$h'' : (n+1)H \to n+k+1$$

of bundles over RP^n. If h is non-singular then h" is injective and the cokernel of h" is a k-plane bundle E such that $E \oplus (n+1)H$ is trivial. Hence E is a normal bundle to RP^n, by (2.1), and so (2.6) follows from Hirsch's theorem.

Whether the converse of (2.6) is true or not appears to be an open question. With the possible exception of $n = 19$ the converse has been shown to be true for $n \leq 23$ through work of Adem [2], [3] and Lam [42], [43], [44], [45]. However Gitler [22] has conjectured that the converse is false for $n = 19, 24, 29, 30$. Some other questions of this type are discussed by Gitler and Lam [24].

I do not know whether, in general, it is possible to give a direct construction for the immersion in (2.6) and avoid the use of Hirsch's theorem. However if

$$h: R^{n+1} \times R^{n+1} \to R^{n+k+1}$$

satisfies a further condition then we can construct an <u>embedding</u> of RP^n as follows. We describe h as <u>symmetric</u> if $h(x,y) = h(y,x)$ for all pairs $x,y \in R^{n+1}$. The example given above has this additional property, when $m = n$ and $t = 2n$. A non-singular bilinear map h determines a map $l: S^n \to S^{n+k}$, where

$$l(x) = h(x,x)\|h(x,x)\|^{-1} (x \in S^n).$$

Note that if $l(x) = l(y)$ $(x,y \in S^n)$ then $h(x,x) = \lambda^2 h(y,y)$ for some real number λ. When h is symmetric this implies that $h(x-\lambda y, x+\lambda y) = 0$, and hence $x = \pm\lambda y$. The map of RP^n determined by l therefore constitutes an embedding and we obtain

<u>Theorem 2.7.</u> <u>If there exists a non-singular symmetric bilinear map of</u> $R^{n+1} \times R^{n+1}$ <u>into</u> R^{n+k+1} <u>then there exists an embedding of</u>

RP^n in R^{n+k}.

The construction on which (2.7) depends is due to Hopf [34], who uses it to illustrate the way topological methods can help to solve problems in algebra. It seems unlikely that the converse of (2.7) is true but I do not know of an example to the contrary.

3. Further results in the real case

At this stage we begin to find it convenient to state conditions in terms of the function $\alpha(n)$ which is defined as the number of non-zero digits in the binary expansion of n. If n is a power of two, for example, then $\alpha(n) = 1$, and conversely.

In this section we describe some special methods for showing that RP^n cannot be immersed when the dimension of the euclidean space is too small. When $n+1$ is a power of two the normal Stiefel-Whitney are all zero, but in this case an entirely different method due to James [38] can be used. This method has been simplified by Adem and Gitler [5] on the following lines.

The Thom spaces of vector bundles over RP^n have been studied by Atiyah [7] who has shown, in particular, that the Thom space of $rH (r = 1,2,\ldots)$ is homeomorphic to RP^{n+r}/RP^{r-1}, the space obtained from RP^{n+r} by collapsing RP^{r-1} to a point. Suppose that we have a normal k-plane bundle to RP^n, for some k. Consider the Thom space T of the corresponding twisted normal bundle. Since T is $(k-1)$-connected the embedding $RP^n \to T$, given by the zero-section, can be deformed into a map which is constant on RP^{k-1}, and hence determines a map $f: RP^n/RP^{k-1} \to T$. From Atiyah's results, using (2.2), it follows that the $(n+1)$-fold suspension $S^{n+1}T$ is homeomorphic to RP^{2n+k+1}/RP^{n+k}. Hence the $(n+1)$-fold suspension of f can be regarded as a map

$$S^{n+1}f:S^{n+1}(RP^n/RP^{k-1}) \to RP^{2n+k+1}/RP^{n+k}.$$

Note that $S^{n+1}f$ is homotopic, for dimensional reasons, to the inclusion of a map

$$g:S^{n+1}(RP^n/RP^{k-1}) \to RP^{2n+1}/RP^{n+k}.$$

Suppose that $n+1$ is a power of two. Then a simple calculation (see [5]) shows that g induces an isomorphism of mod 2 cohomology, and hence that RP^n/RP^{k-1} is S-reducible if RP^{2n+1}/RP^{n+k} is S-reducible. The S-reducibility problem for these spaces has been settled by Adams [1] using the ψ-operations of K-theory, and we at once deduce

<u>Theorem</u> 3.1. <u>Let</u> $n = 2^r-1$, <u>where</u> $r \geq 4$. <u>Then</u> $RP^n \not\subseteq R^{2n-q}$, <u>where</u>

$$q = 2r (r \equiv 1,2 \bmod 4),$$

$$q = 2r+1 (r \equiv 0 \bmod 4),$$

$$q = 2r+2 (r \equiv 3 \bmod 4).$$

Efforts have been made to apply a similar method to other cases but this appears to be an isolated result. The conjecture that (3.1) is a best possible result has been established by Gitler and Mahowald [27] who used Postnikov methods to prove

<u>Theorem</u> 3.2. <u>If</u> n <u>and</u> q <u>are as in</u> (3.1) <u>then</u> $RP^n \subseteq R^{2n-q+1}$.

Of course the cases $n = 3, 7$ are already dealt with in 1.

Various other immersion results have been obtained by Postnikov methods including

<u>Theorem</u> 3.3. <u>If</u> $n \equiv 0 \bmod 4$ <u>and</u> n <u>is not a power of two then</u> $RP^n \subseteq R^{2n-5}$. <u>If</u> $n \equiv 4 \bmod 8$ <u>and</u> $n-4$ <u>is not a power of two then</u> $RP^n \subseteq R^{2n-7}$. <u>If</u> $n \equiv 0 \bmod 8$ <u>and</u> n <u>is not a power of two then</u> $RP^n \subseteq R^{2n-9}$.

<u>Theorem</u> 3.4. <u>If</u> n <u>is odd and</u> $\alpha(n) > 3$ <u>then</u> $RP^n \subseteq R^{2n-8}$. <u>If</u> $n \equiv 3 \bmod 4$ <u>and</u> $\alpha(n) > 5$ <u>then</u> $RP^n \subseteq R^{2n-9}$.

The first part of (3.3) is due to Gitler and Mahowald [25]. The remainder of (3.3) and (3.4) in case $n \equiv 1 \bmod 4$ is due to Randall [65], while (3.4) in case $n \equiv 3 \bmod 4$ is due to Johnson [40]. Since all these results turn on somewhat complicated calculations it is unfortunately impossible to say much about them here. However the general approach is dealt with in Prof. Gitler's lectures in this same volume.

Returning to negative results, much useful information is contained in

<u>Theorem</u> 3.5. <u>If</u> n <u>is of the form</u> $2^r+2^s+1 (r > s \geq 2)$ <u>or</u> $2^r+2 (r > 2)$ <u>then</u> $RP^n \not\subseteq R^{2n-5}$. <u>If</u> n <u>is of the form</u> $2^r+4 (r > 2)$ <u>then</u> $RP^n \not\subseteq R^{2n-7}$.

The first part of (3.5) was established by Adem and Gitler in [5] by means of secondary operations. The case $n = 2^r+2$ was established by Baum and Browder [10] as an application of their analysis of the cohomology structure of the projective orthogonal group. Subsequently Gitler and Handel [23] extended this analysis to the projective Stiefel manifold. This enters into the immersion problem as follows. Consider the Stiefel manifold $V_{n+k+1,n+1}$ of orthonormal n-frames of tangents to S^{n+k}, with the involution given by transfering the n-frame to the antipodal point on S^{n+k} and simultaneously changing the sign of each of the n vectors. By factoring out this action we obtain the projective Stiefel manifold $U_{n+k+1,n+1}$, with RP^{n+k} embedded in the obvious way. It follows from (2.3) that RP^n can be immersed in R^{n+k} if and only $U_{n+k+1,n+1}$ admits a cross-section over RP^n. This implies conditions on the cohomology from which Gitler [21] obtains (3.5).

4. Further negative results

The main purpose of this section is to give some account of the information to be obtained from the integrality theorems of Atiyah, Hirzebruch, Mayer, Sanderson and Schwarzenberger. These are primarily designed for manifolds without 2-torsion; nothing useful is obtained in the real case.

Consider the integral cohomology of a manifold X, as before. The (tangential) Pontrjagin classes of X are denoted by $p_i(X) \in H^{4i}(X)$ $(i = 1,2,...)$ in the usual way, and the normal Pontrjagin classes by $\bar{p}_i(X) \in H^{4i}(X)$. Recall that the square of the Euler class of a 2t-plane bundle, where $t \geq 1$, is equal to the Pontrjagin class in dimension 4t. Hence $\bar{p}_t(X)$ is a perfect square if $X \subseteq R^{m+2t}$, where $m = \dim X$. In this way it can be shown (private communication from Prof. Mahowald) that the Cayley projective plane cannot be immersed in R^{24}; we recall that this 16-manifold can be embedded in R^{25}. The same argument shows that HP^2 cannot be immersed in R^{12}. More generally we have

<u>Theorem</u> 4.1. <u>If</u> n <u>is a power of two then</u> $HP^n \not\subseteq R^{8n-4}$.

In case $n > 2$ this result is proved by Mahowald and Peterson [55] using secondary cohomology operations.

The integrality theorem of Atiyah and Hirzebruch [9] applies in case X is an orientable 2m-manifold. Certain polynomials in the Pontrjagin classes $p_1(X),\ldots,p_m(X)$ are defined, with rational coefficients. The theorem asserts that if $X \subset R^{2m+2l}$ then each of the rational numbers obtained by evaluating these classes on X is the quotient of an integer by 2^{m+l-1}. In case X satisfies certain further conditions we can replace the divisor by 2^{m+l-2}. The polynomials arise from consideration of spin representations; we omit the details. A useful lemma of Sanderson and Schwarzenberger [69] shows that exactly the same condition holds on the alternative hypothesis that $X \subseteq R^{2m+2l-1}$. A number of variants of the basic integrality theorem have been established by Mayer and Schwarzenberger [60] and by Mayer [58], [59]. If X is a complex or quaternionic projective space the conclusions reached are

<u>Theorem</u> 4.2. <u>If</u> $n \geq 2$ <u>then</u> CP^n <u>cannot be immersed in</u> $R^{4n-2\alpha(n)-1}$, <u>or embedded in</u> $R^{4n-2\alpha(n)-2}$. <u>Also</u> HP^n <u>cannot be immersed in</u> $R^{8n-2\alpha(n)-3}$ <u>or embedded in</u> $R^{8n-2\alpha(n)-2}$.

<u>Theorem</u> 4.3. <u>Let</u> n <u>be even</u>. <u>If</u> $\alpha(n) \equiv 1 \bmod 4$ <u>then</u> $CP^n \not\subseteq R^{4n-2\alpha(n)}$. <u>If</u> $\alpha(n) \equiv 2$ <u>or</u> $3 \bmod 4$ <u>then</u> $CP^n \not\subseteq R^{4n-2\alpha(n)+1}$. <u>If</u> $\alpha(n) \equiv 3 \bmod 4$ <u>then</u> $CP^n \not\subseteq R^{4n-2\alpha(n)+2}$.

<u>Theorem</u> 4.4. <u>Let</u> $n \geq 1$. <u>If</u> $\alpha(n) \equiv 2 \bmod 4$ <u>then</u> $HP^n \not\subseteq R^{8n-2\alpha(n)-2}$. <u>If</u> $\alpha(n) \equiv 0$ <u>or</u> $3 \bmod 4$ <u>then</u> $HP^n \not\subseteq R^{8n-2\alpha(n)-1}$.

If $\alpha(n) \equiv 0 \bmod 4$ *then* $HP^n \not\subset R^{8n-2\alpha(n)}$.

Some of these results have been obtained by Steer [70] using a rather different method, and others by Feder and Segal [19] (correcting an error in Feder [17]) using yet another method. In all cases, however, the fundamental process is K-theoretic.

In the real case there are a few additional non-embedding results which we collect together as

Theorem 4.5. *If* $n-1$ *is a power of two then* $RP^n \not\subset R^{2n-2}$. *If* $n-5$ *is a power of two and* $n \geq 21$ *then* $RP^n \not\subset R^{2n-4}$. *If* $n-3$ *is of the form* $2^r+2^s (r > s \geq 2)$ *then* $RP^n \not\subset R^{2n-8}$.

The first of these three results has been established independently by Levine [46] and Mahowald [49]. The remaining results are proved by Adem, Gitler and Mahowald [6], using secondary cohomology operations.

5. Further positive results

We can regard the sphere S^{m+n+1} as the join $S^m * S^n$ of S^m with S^n. Points of the join are represented by triples

$$(x,y,t) \quad (x \in S^m,\ y \in S^n, t \in I)$$

subject to the usual identifications. To obtain RP^{m+n+1} we identify (x,y,t) with $(-x,-y,t)$. In RP^{m+n+1} the triples where $t = 0$ form a subspace RP^m, while those where $t = 1$ form a (disjoint) subspace RP^n. We denote the Hopf line bundle over RP^q by $H_q (q = m,n)$. The subspace of RP^{m+n+1} where $t \leq \frac{1}{2}$ can be identified with the disc-bundle of $(n+1)H_m$; the subspace where

$t \geq \frac{1}{2}$ with the disc-bundle of $(m+1)H_n$. Thus we can construct RP^{m+n+1} from these two disc-bundles by identifying the total spaces of their associated sphere-bundles in an appropriate fashion. In the literature (see Atiyah [7], Epstein and Schwarzenberger [16], Milgram [61] and Steer [71]) the properties of this construction are exploite in various ways. Following Milgram [61] we refer to it as the projective join.

The projective join can be used to construct immersions of RP^{m+n+1} as follows. First choose immersions of the disc-bundles of $(n+1)H_m$ and $(m+1)H_n$. Next consider the restrictions of these immersions to the total spaces of the associated sphere-bundles. Transfer one of these to the other space, using the homeomorphism determined by the intrinsic join construction. Then we have two immersions of (say) the sphere-bundle associated with $(n+1)H_m$, and i these can be deformed into each other by a regular homotopy then an immersion of RP^{m+n+1} can be obtained. A similar method can be used to construct embeddings. Complex and quaternionic projective spaces can be treated similarly.

There are two basic ways of obtaining normal bundles, and hence immersions. One is to construct a monomorphism of the tangent bundle of appropriate dimension and take the cokernel. The other is to construct an epimorphism of the trivial bundle onto the cotangent bundle and take the kernel. Milgram [61] adopts the second approach and constructs "spanning sets" by methods of linear algebra. These originate in low dimensions, where they can be written down explicitly, and are combined by a process based on the projective join. Using this method we obtain

Theorem 5.1. *Let* $n > 7$. *If* $n \equiv 0$ *or* $1 \bmod 8$ *then* $RP^n \subseteq R^{2n-\alpha(n)}$. *If* $n \equiv 3$ *or* $5 \bmod 8$ *then* $RP^n \subseteq R^{2n-\alpha(n)-1}$. *If* $n \equiv 7 \bmod 8$ *then* $RP^n \subseteq R^{2n-\alpha(n)-4}$.

This result is due to Milgram [61] except for the case $n \equiv 7 \bmod 8$ which is obtained by Lam [44]. Further information is obtained by Nussbaum [64] (the general case), by Lam [42] (the case $n = 12$) and Adem [2] (the case $n = 20$) who prove

Theorem 5.2. *If* n *is of the form* 2^r+4, *where* $r \geq 3$, *then* $RP^n \subseteq R^{2n-6}$.

Steer [71] has shown that Milgram's procedure can also be used for the construction of embeddings. Recently Rees [66] has shown that RP^{15} can be embedded topologically in R^{23}. This result can be used to start one of Steer's inductive processes and then the result stated in [71] can be improved to

Theorem 5.3. *If* $n > 7$ *and* $n \not\equiv 1 \bmod 8$ *then* $RP^n \subset R^{2n-\alpha(n)}$. *If* $n > 15$ *and* $n \equiv 7 \bmod 8$ *then* $RP^n \subset R^{2n-\alpha(n)-3}$.

Milgram and Rees [62] have developed the theory in another direction and this has enabled Rees [66], with the help of the Browder embedding theorem, to prove

Theorem 5.4. *If* $n \equiv 1 \bmod 8$ *and* $\alpha(n) \leq 8$ *then* $RP^n \subset R^{2n-\alpha(n)+1}$. *If* $n \equiv 3$ *or* $5 \bmod 8$ *and* $\alpha(n) \leq 6$ *then* $RP^n \subset RP^{2n-\alpha(n)-1}$.

Theorem 5.5. *If* $n \equiv 7 \bmod 8$ *and* $n \geq 23$ *then* $RP^n \subset R^{2n-7}$. *If* $n \equiv 7 \bmod 32$ *and* $n \geq 39$ *then* $RP^n \subset R^{2n-8}$.

In the complex case Steer [71] proves

<u>Theorem</u> 5.6. <u>If</u> $n \geq 2$ <u>then</u> $CP^n \subset R^{4n-\alpha(n)}$. <u>If</u> n <u>is odd and</u> $n-1$ <u>is not a power of two then</u> $CP^n \subset R^{4n-\alpha(n)-1}$.

The existence of immersions under precisely the same conditions was established by Milgram [61]. Finally Milgram [61] (the immersion case) and Steer [71] (the embedding case) have proved

<u>Theorem</u> 5.7. <u>If</u> $n \geq 2$ <u>then</u> HP^n <u>can be immersed in</u> $R^{8n-\alpha(n)+3}$, <u>and embedded in</u> $R^{8n-\alpha(n)+4}$.

For low values of $\alpha(n)$ these general results can be improved as follows

<u>Theorem</u> 5.8. <u>If</u> $n > 3$ <u>and</u> $\alpha(n) > 1$ <u>then</u> $CP^n \subset R^{4n-2}$. <u>If</u> n <u>is odd and</u> $\alpha(n) > 2$ <u>then</u> $CP^n \subseteq R^{4n-5}$. <u>If</u> $n \equiv 3 \bmod 4$ <u>then</u> $CP^n \subseteq R^{4n-6}$.

<u>Theorem</u> 5.9. <u>If</u> $n > 1$ <u>and</u> $\alpha(n) > 1$ <u>then</u> $HP^n \subset R^{8n-4}$ <u>and</u> $HP^n \subseteq R^{8n-5}$. <u>If</u> $\alpha(n) > 2$ <u>then</u> $HP^n \subset R^{8n-5}$ <u>and</u> $HP^n \subseteq R^{8n-6}$.

In both theorems the first assertions are due to Sanderson [68]. The second assertion in (5.8) is proved by Randall [65], using Postnikov methods. An independent, but similar, proof has been given by Johnson [40], who also established the second assertions in (5.9). The third assertion in (5.8) is due to Steer [73]. By applying (1.5) to these results we obtain information concerning the real case, including some of the results mentioned previously.

For low values of $\alpha(n)$ the general results on embeddings can also be improved by a theorem of Mahowald [51] as follows

Theorem 5.10. *If* $n > 3$ *and* $n \equiv 3 \bmod 4$ *then* $RP^n \subset R^{2n-2}$. *If* n *is even and neither* n *nor* $n-2$ *is a power of two then* $RP^n \subset R^{2n-3}$. *If* $n \equiv 1 \bmod 4$ *and* $n-1$ *is not a power of two then* $RP^n \subset R^{2n-3}$.

Finally we remark that Rees [66] has shown that $RP^{14} \subset R^{23}$ and has also shown that there exist piecewise-linear embeddings of RP^7 in R^{10} and of RP^{15} in R^{23}.

. A few "best possible" immersions

The various results we have mentioned determine, in a range of cases, the least value of k such that $RP^n \subseteq R^k$. This "best possible" dimension is given by (3.1) and (3.2) in case n is of the form $2^r-1 (r \geq 4)$, and by the following table when $n = 2^r,\dots,2^r+7$ ($r \geq 3$)

n	k	reference
2^r	2n-1	(1.6), (1.2)
2^r+1	2n-3	(1.6), (1.2)
2^r+2	2n-4	(1.6), (3.5)
2^r+3	2n-6	(1.6), (3.5)
2^r+4	2n-6	(5.2), (3.5)
2^r+5	2n-4	(5.1), (3.5)
2^r+6	2n-6	(5.1), (3.5)
2^r+7	2n-8	(5.1), (3.5)

REFERENCES

[1] Adams, J.F., Vector fields on spheres, Ann. of Math. 75(1962), 603-632.

[2] Adem, J., Some immersions associated with bilinear maps, Bol. Soc. Mat. Mex. 13(1968), 95-104.

[3] Adem, J., On non-singular bilinear maps, Springer-Verlag Lectur Notes in Mathematics, 168(1970), 11-24.

[4] Adem, J. and Gitler, S., Secondary characteristic classes and the immersion problem, Bol. Soc. Mat. Mex. 8(1963), 53-78.

[5] Adem, J. and Gitler, S., Non-immersion theorems for real projective spaces, Bol. Soc. Mat. Mex. 9(1964), 37-50

[6] Adem, J., Gitler, S., and Mahowald, M., Embedding and immersion of projective spaces, Bol. Soc. Mat. Mex. 10(1966), 8 88.

[7] Atiyah, M.F., Thom complexes, Proc. London Math. Soc. (3) 11(1961), 291-310.

[8] Atiyah, M.F., Immersions and embeddings of manifolds, Topology 1(1962), 125-132.

[9] Atiyah, M.F. and Hirzebruch, F., Quelques théorèmes de nonplongement pour les variétés différentiables, Bull Soc. Math. France 87(1959), 383-396.

[10] Baum, P. and Browder, W., The cohomology of quotients of classical groups, Topology 3(1965), 305-336.

[11] Behrend, F., Uber systeme reeller algebraischer gleichungen, Compositio Math. 7(1939), 1-19.

[12] Borel, A. and Hirzebruch, F., Characteristic classes and

homogeneous spaces I, II, III, Amer. J. of Math. 80 (1958), 458-538; 81(1959), 315-382; 82(1960), 491-504.

[13] Browder, W., Embedding smooth manifolds, Proc. I.C.M. Moscow (1966), 712-719.

[14] Chern, Shiing-Shen, On the multiplication in the characteristic ring of a sphere-bundle, Ann. of Math. 49(1948), 362-372.

[15] Davis, D., Generalized homology and the generalized vector field problem (preprint).

[16] Epstein, D.B.A. and Schwarzenberger, R.L.E., Imbeddings of real projective spaces, Ann. of Math. 76(1962), 180-184.

[17] Feder, S., Non-immersion theorems for complex and quaternionic projective spaces, Bol. Soc. Mat. Mex. 11(1966), 62-67.

[18] Feder, S., The reduced symmetric product of a projective space and the embedding problem, Bol. Soc. Mat. Mex. 12(1967) 76-78.

[19] Feder, S. and Segal, D.M., Immersions and embeddings of projective spaces, Bull. Amer. Math. Soc. (to appear).

[20] Ginsburg, M., Some immersions of projective space in euclidean space, Topology 2(1963), 69-71.

[21] Gitler, S., The projective Stiefel manifolds II. Applications, Topology 7(1968), 47-53.

[22] Gitler, S., Immersion and embedding of manifolds, Proc. Summer School Algebraic Topology Madison 1970 (to appear).

[23] Gitler, S. & Handel, D., The projective Stiefel manifolds I, Topology 7(1967), 39-46.

[24] Gitler, S. and Lam, K.Y., The generalized vector field problem

and bilinear maps, Bol. Soc. Mat. Mex. 14(1969), 65-69

[25] Gitler, S. and Mahowald, M., The geometric dimension of real stable vector bundles, Bol. Soc. Mat. Mex. 11(1966), 85-107.

[26] Gitler, S. and Mahowald, M., The immersion of manifolds, Bull. Amer. Math. Soc. 73(1967), 696-700.

[27] Gitler, S. y Mahowald, M., Some immersions of real projective spaces, Bol. Soc. Mat. Mex. 14(1969), 9-21.

[28] Haefliger, A., Differentiable embeddings, Bull. Amer. Math. Soc. 67(1961), 109-112.

[29] Haefliger, A., Plongements différentiables dans le domaine stable, Comment. Math. Helv. 37(1962), 155-176.

[30] Haefliger, A. and Hirsch, M., Immersions in the stable range, Ann. of Math. 75(1962), 231-241.

[31] Handel, D., On the normal bundle of an embedding, Topology 6(1967), 65-68.

[32] Handel, D., An embedding theorem for real projective spaces, Topology 7(1968), 125-130.

[33] Hirsch, M.W., Immersion of manifolds, Trans. Amer. Math. Soc. 93(1959), 242-276.

[34] Hopf, H., Systeme symmetrischer Bilinearformen und euklidische Modelle der projecktiven Raüme, Vjschr. naturf. Ges. Zurich 85(1940), 165-177.

[35] Hopf, H., Ein topologischer Beitrag zur reellen Algebra, Comment. Math. Helv. 13(1940-41), 219-239.

[36] James, I.M., Embeddings of real projective spaces, Proc. Cambridge Phil. Soc. 54(1958), 555-557.

[37] James, I.M., Some embeddings of projective spaces, Proc. Cambridge Phil. Soc. 55(1959), 294-298.

[38] James, I.M., On the immersion problem for real projective spaces, Bull. Amer. Math. Soc. 69(1963), 231-238.

[39] James, I.M., Euclidean models of projective spaces, Bull. London Math. Soc. 3(1971), 257-276.

[40] Johnson, A.L., The immersion and embedding of complex and quaternionic projective spaces (to appear).

[41] Kobayashi, T., On the odd order non-singular immersions of real projective spaces, J. Sci. Hiroshima Univ. 33(1969), 197-207.

[42] Lam, K.Y., Construction of non-singular bilinear maps, Topology 6(1967), 423-426.

[43] Lam, K.Y., On bilinear and skew linear maps that are non-singular maps, Bol. Soc. Mat. Mex. 13(1968), 88-94.

[44] Lam, K.Y., Construction of some non-singular bilinear maps, Bol. Soc. Mat. Mex. 13(1968), 88-94.

[45] Lam, K.Y., Sectioning vector bundles over real projective spaces (to appear).

[46] Levine, J., Embedding and immersion of real projective spaces, Proc. Amer. Math. Soc. 14(1963), 801-803.

[47] Levine, J., On differentiable embeddings of simply connected manifolds, Bull. Amer. Math. Soc. 69(1963), 806-809.

[48] Levine, J., On obstructions in sphere-bundles and immersions of manifolds, Trans. Amer. Math. Soc. 109(1963), 420-429.

[49] Mahowald, M., On the embeddability of the real projective spaces, Proc. Amer. Math. Soc. 13(1962), 763-764.

[50] Mahowald, M., On the normal bundle of a manifold, Pacific J. of Math. 14(1964), 1335-1341.

[51] Mahowald, M., On obstruction theory in orientable fibre bundles Trans. Amer. Math. Soc. 110(1964), 315-349.

[52] Mahowald, M., On embedding manifolds which are bundles over spheres, Proc. Amer. Math. Soc. 15(1964) 579-583.

[53] Mahowald, M. and Milgram, R.J., Embedding projective spaces, Bull. Amer. Math. Soc. 73(1967), 644-646.

[54] Mahowald, M. y Milgram, R.J., Embedding real projective spaces, Ann. of Math. 87(1968), 411-422.

[55] Mahowald, M.E. and Peterson, F.P., Secondary cohomology operations on the Thom class, Topology 2(1963), 367-377.

[56] Massey, W.S., On the cohomology ring of a sphere bundle, J. Mat Mech. 7(1958), 265-290.

[57] Massey, W.S., On the embeddability of the real projective space in Euclidean space, Pacific J. of Math. 9(1959), 783-789.

[58] Mayer, K.H., Elliptische Differentialoperatoren und Ganzahligkeitssätze für Charakteristische Zahlen, Topology 4(1965), 295-313.

[59] Mayer, K.H., Ganzahligkeitssätze und Immersionem von Sphärenbündeln über sphären, Math. Annalen 174(1967), 195-202.

[60] Mayer, K.H. and Schwarzenberger, R.L.E., Non-embedding theorems for Y-spaces, Proc. Cambridge Phil. Soc. 63(1967), 601-612.

[61] Milgram, R.J., Immersing projective spaces, Ann. of Math. 85 (1967), 473-482.

[62] Milgram, R.J. and Rees, E., On the normal bundle to an embedding, Topology 10(1971), 299-308.

[63] Nakagawa, R., Embeddings of projective spaces and lens spaces, Sci. Rep. Tokyo Kyoiku Daigaku A9(1967), 170-175.

[64] Nussbaum, F., Non-orientable obstruction theory, Thesis, Northwestern University 1970.

[65] Randall, A.D., Some immersion theorems for projective spaces, Trans. Amer. Math. Soc. 147(1970), 135-151.

[66] Rees, E., Embeddings of real projective spaces, Topology 10(1971), 309-312.

[67] Sanderson, B.J., A non-immersion theorem for real projective space, Topology 2(1963), 209-211.

[68] Sanderson, B.J., Immersions and embeddings of projective spaces, Proc. London Math. Soc. (3) 14(1964), 137-153.

[69] Sanderson, B.J. and Schwarzenberger, R.L.E., Non-immersion theorems for differentiable manifolds, Proc. Cambridge Phil. Soc. 59(1963), 319-322.

[70] Steer, B., Une interpretation géométrique des nombres de Radon-Hurwitz, Annales de l'Institut Fourier, 17(1967), 209-218.

[71] Steer, B., On the embeddings of projective spaces in euclidean space, Proc. London Math. Soc. 21(1970), 489-501.

[72] Steer, B., On immersing complex projective (4k+3)-space in Euclidean space, Quart J. Math. Oxford 22(1971), 347-351.

[73] Suzuki, H., Bounds for dimensions of odd order nonsingular immersions of RP^n, Trans. Amer. Math. Soc. 121(1966), 269-275.

[74] Thom, R., Espaces fibrés en sphères et carrés de Steenrod, Ann. Sci. Ecole Norm. Sup. (3) 69(1952), 109-182.

[75] Thomas, E., Vector fields on manifolds, Bull. Amer. Math. Soc. 75(1969), 643-683.

[76] Tomonaga, Y., Differentiable embedding and cobordism of orientable manifolds, Tohuku Math. J. 14(1962), 15-23

[77] Tomonaga,Y., Certain problems of differentiable imbedding, Proc Amer. Math. Soc. 14(1963), 787-789.

[78] Vranceanu, G., Sur le plongement des espaces projectifs, C.R. Acad. Sci. Paris, 251(1960), 192-193.

[79] Watabe, T., Imbedding and immersion of projective spaces, Science Reports Niigata Univ. A (1965), 11-16.

[80] Whitney, H., The self-intersections of a smooth n-manifold in 2n-space, Ann. of Math. 45(1944), 220-246.

[81] Whitney, H., The singularities of a smooth n-manifold in (2n-1)-space, Ann. of Math. 45(1944), 247-293.

[82] Wu Wen-Tsun, A theory of imbedding immersion and isotopy of polytopes in a euclidean space, Science Press 1965 (Peking).

[83] Yo Ging-Tzung, Cohomology of the projective bundle of a manifold and its applications to immersions, Sci. Sinica 14 (1965), 959-963.

[84] Yosida, T., On the vector bundle $m\xi_n$ over real projective spaces, J. Sci. Hiroshima Univ. Ser. A-I Math. 32(1968) 5-16.

PART II: BUNDLES WITH SPECIAL STRUCTURE

1. Introduction

Let E be an orthogonal sphere-bundle over a base space B. We describe a fibre-preserving map $f:E \to E$ as an A-structure if fx is orthogonal to x for all $x \in E$. We describe a fibre-preserving homotopy $g_t:E \to E (t \in I)$ as a homotopy-symmetry if $g_0x = x$ and $g_1x = -x$ for all $x \in E$. We observe that an A-structure f determines a homotopy-symmetry g_t, where

(1.1) $$g_t x = x \cos \pi t + f(x) \sin \pi t.$$

Among the fibre preserving maps of E into itself are those which are orthogonal, in the sense that each fibre is transformed into itself by an orthogonal transformation. Hence the notion of orthogonal A-structure is defined. Also, by extending the term to a family of maps, so is the notion of orthogonal homotopy-symmetry. Note that g_t, in (1.1), is orthogonal if f is orthogonal.

One condition needs to be satisfied before E can have any of these four properties. Suppose that E is an (n-1)-sphere bundle. The antipodal map on the (n-1)-sphere has degree $(-1)^n$. Hence n is even if E admits a homotopy-symmetry, and a fortiori if E admits an A-structure. Consequently we assume, henceforth, that the fibres of E are spheres of odd dimension.

Let ξ be the n-plane bundle associated with E, so that $E = S(\xi)$. Let λ denote the Hopf line bundle over S^1, regarded as the real projective line. Consider the n-plane bundles $\xi \otimes \lambda$ and

$\xi \otimes 1$, over $B \times S^1$. We identify $\xi \otimes \lambda | B \vee S^1$ with $\xi \otimes 1 | B \vee S^1$, in the obvious way, and prove

Theorem 1.2. *The sphere-bundle* $S(\xi)$ *admits a homotopy-symmetry if and only if there exists a fibre-homotopy equivalence between* $S(\xi \otimes \lambda)$ *and* $S(\xi \otimes 1)$, *extending the identity on* $B \vee S^1$.

Hence and from the formula (see [20]) for the (mod 2) Stiefel-Whitney classes of the tensor product we obtain

Corollary 1.3. *In the case of a homotopy-symmetric sphere-bundle all the odd-dimensional Stiefel-Whitney classes are zero; in particular the bundle is orientable.*

The conclusion of (1.3) is true, a fortiori, if the bundle admits an A-structure. In this form the corollary has been proved in [14] by a different method involving the relative Stiefel-Whitney classes of Kervaire [17]. Of course (1.3) is well-known in case the sphere-bundle admits an orthogonal A-structure since then the existence of Chern classes shows that the even-dimensional Stiefel-Whitney classes are the reductions of integral classes, and hence the conclusion follows from consideration of the appropriate coefficient exact sequence. By the reverse of this argument we deduce from (1.3), as it stands, that the even-dimensional Stiefel-Whitney classes of a homotopy-symmetric sphere-bundle are the reductions of mod 4 classes and examples can be given (see [15]) to show that they are not necessarily the reductions of integral classes.

To prove (1.2) we represent points of S^1 by complex numbers of unit modulus, in the usual way. Let $p: E \to B$ denote the projection, where $E = S(\xi)$. Consider the space $\widetilde{E}$ formed from $E \times S^1$ by identifying (x,z) with $(-x,-z)$ for all $x \in E$, $z \in S^1$. We fibre $\widetilde{E}$ over $B \times S^1$ with projection $\widetilde{p}$ given by $\widetilde{p}(x,z) = (x,z^2)$, and observe that $\widetilde{E} = S(\xi \otimes \lambda)$. Suppose that we have a homotopy-symmetry $g_t : E \to E$. Then a map $h : E \times S \to E$ is defined by $h(x, \exp \pi t) = g_t x = h(-x, -\exp \pi t)$, for $0 \leq t \leq 1$. Furthermore h determines a map $k : \widetilde{E} \to E$ such that $pk = l\widetilde{p}$, where $l : B \times S^1 \to B$ denotes the left projection. Let $k' : \widetilde{E} \to E \times S^1 = l^* E$ be given by $k'(x,z) = (k(x,z), z^2)$. Then a standard argument (see [6]) shows that k' is a fibre-homotopy equivalence, rel. $B \vee S^1$. This establishes (1.2) in one direction and a straightforward reversal of the reasoning completes the proof. Notice, moreover, that k' is orthogonal if and only if g_t is orthogonal, so that we also obtain

<u>Theorem</u> 1.4. <u>The sphere-bundle</u> $S(\xi)$ <u>admits an orthogonal homotopy-symmetry if and only if there exists an isomorphism between</u> $\xi \otimes \lambda$ <u>and</u> $\xi \otimes 1$, <u>extending the identity on</u> $B \vee S^1$.

Of course these results give useful necessary (but in general not sufficient) conditions for the existence of ordinary or orthogonal A-structure. Moreover, the existence of a homotopy-symmetry (resp. orthogonal homotopy-symmetry) implies the existence of an A-structure (resp. orthogonal A-structure) when the sphere-bundle concerned is stable, in an appropriate sense. This will be shown in the next section. Of course (1.2) and (1.4) can, with some loss of strength, be reformulated in terms of K-theory. For this purpose, however, it is possible to use other methods as indicated by James [10], extended by Becker [4] and further extended by Woodward (in preparation).

2. In the stable range

As before let E be an orthogonal (n-1)-sphere bundle over E, with n even. Consider the group O_n of orthogonal transformations and the subspace O'_n of orthogonal skew-symmetric transformations. Make O_n act on O'_n by conjugation. Recall (see [19]) the well-known

Theorem 2.1. *The sphere-bundle E admits orthogonal A-structure if, and only if, the associated bundle with fibre O'_n admits a cross-section.*

Consider also the space O''_n of paths in O_n from e to -e, where e denotes the identity transformation. We make O_n act on itself by conjugation (the adjoint action) and transmit this to O''_n in the obvious way. By modification of an appropriate proof of (2.1) we obtain

Theorem 2.2. *The sphere-bundle E admits an orthogonal homotopy-symmetry if, and only if, the associated bundle with fibre O''_n admits a cross-section.*

An O_n-equivariant map $h:O'_n \to O''_n$ is given by

$$h(a) = e \cos \pi t + a \sin \pi t \quad (a \in O'_n, t \in I).$$

Consider the corresponding map of the associated bundle with fibre O'_n into the associated bundle with fibre O''_n. It follows from the main theorem of Bott [5] that

$$h_*:\pi_r(O'_n) \longrightarrow \pi_r(O''_n) \approx \pi_{r+1}(O_n)$$

is bijective for $r \leq n-3$. Suppose that B is a locally-finite CW-complex of finite (cohomological) dimension. If $\dim B \leq n-2$ then, using classical obstruction theory, it follows that the associated bundle with fibre O'_n admits a cross-section if the associated bundle with fibre O''_n admits a cross-section. Thus we obtain

Theorem 2.3. *Suppose that* $\dim B \leq n-2$. *Then* E *admits an orthogonal* A-*structure if* (*and only if*) E *admits an orthogonal homotopy-symmetry*.

In the ordinary case, the corresponding result is

Theorem 2.4. *Suppose that* $\dim B \leq n-4$. *Then* E *admits an* A-*structure if* (*and only if*) E *admits a homotopy-symmetry*.

Perhaps the easiest way to prove (2.4) is as follows. Let $p:E \longrightarrow B$ denote the given fibration. Let E' now denote the space of pairs (x,y), where $x, y \in E$, such that $px = py$ and such that x is orthogonal to y. We fibre E' over E with projection p' given by $p'(x,y) = x$. An A-structure f on E determines a cross-section $f':E \longrightarrow E'$, where $f'x = (x,fx)$, and conversely a cross-section determines an A-structure.

Let E" denote the space of paths λ in E such that $p\lambda$ is stationary in B and such that $\lambda(0) = -\lambda(1)$. We fibre E" over E with projection p" given by $p''\lambda = \lambda(0)$. Then $p''h = p'$, where $h:E' \longrightarrow E''$ is the map defined by

$$h(x,y)(t) = x \cos \pi t + y \sin \pi t \quad (t \in I).$$

A homotopy-symmetry $g_t:E \to E$ determines a cross-section $f'':E \to E''$, where $f''(x)(t) = g_t(x)$, and conversely.

Consider the fibre S^{n-1} of the original fibration $p:E \to B$ which contains the basepoint $e \in E$. The fibre of $p':E' \to E$ over can be identified with S^{n-2}, the equator in the hyperplane orthogon to E. Also the corresponding fibre of $p'':E'' \to E$ can be identifie with ΩS^{n-1}, the space of paths in S^{n-1} from e to $-e$, so that maps $x \in S^{n-2}$ into the great semicircle through x. Hence it follows from the Freudenthal suspension theorem that

$$h_*:\pi_r(S^{n-2}) \to \pi_r(\Omega S^{n-1}) = \pi_{r+1}(S^{n-1})$$

is bijective for $r \leq 2n-6$. By hypothesis we have $\dim E = \dim B + n-1 \leq 2n-5$, so it follows from classical obstruction theory that $p':E \to E$ admits a cross-section if $p'':E'' \to E$ admits a cross section. This proves (2.4). Notice that the "stable range" for (2. includes two dimensions which the "stable range" for (2.4) does not.

In view of (1.3) let us now assume that E is oriented, i.e. that the structural group is reduced to the special orthogonal group $R_n \subset O_n$. Since the space O_n' has (exactly) two components it follow that the associated bundle with fibre O_n' has (exactly) two components. The image of a cross-section of this bundle may lie in either component; thus we can distinguish two types of orthogonal A-structure. Consider the adjoint action of an element of $O_n - R_n$, i.e. an improper orthogonal transformation. This action transforms E into an oriented bundle $\bar{E}$, and at the same time transforms an orthogonal A-structure f on E into an orthogonal A-structure $\bar{f}$ on $\bar{E}$. However, f and $\bar{f}$ are of different types (in the above sense). The situation is precisely similar for ordinary A-structure when $n \geq 6$, and for both ordinary and orthogonal homotopy-symmetry when $n \geq 4$. When $n = 4$ however, it is necessary (as observed by Hopf) to distinguish not just two but an infinity of types of A-structure.

To clarify the situation we return to the proof of (2.4), with B a point-space, so that $E = S^{n-1}$. Write $E' = V'_n$ (instead of the usual notation for the Stiefel manifold) and write $E'' = V''_n$. Recall that S^{n-2} is the fibre of $p':V'_n \to S^{n-1}$ and $\Omega(S^{n-1})$ is the fibre of $p'':V''_n \to S^{n-1}$. If γ is any element (such as a generator) of $\pi_{n-1}(S^{n-1})$ then the set $\pi_{n-1}(S^{n-2})$ is equivalent to the subset $p'^{-1}_*\gamma \subset \pi_{n-1}(V'_n)$, also the set $\pi_{n-1}(\Omega S^{n-1})$ is equivalent to the subset $p''^{-1}_*\gamma \subset \pi_{n-1}(V''_n)$. It follows that p' admits two classes of cross-section when $n \geq 6$, and an infinite number when $n = 4$. Also p'' admits two classes of cross-section when $n \geq 4$.

Now let E, as before, be an oriented (n-1)-sphere bundle over B. An A-structure $f:E \to E$ determines an A-structure on S^{n-1}, by restriction to one of the fibres, and hence an element of $\pi_{n-1}(V'_n)$. This element does not depend on the choice of fibre and will be denoted by $\alpha(f)$. Similarly a homotopy-symmetry $g_t:E \to E$ determines an element $\beta(g_t) \in \pi_{n-1}(V''_n)$. We refer to $\alpha(f)$ as the type of f, and to $\beta(g_t)$ as the <u>type</u> of g_t. There is an obvious relationship between these two elements in case g_t is determined by f as in (1.1).

Note that if $f:E \to E$ is an A-structure then so is $-f$, where $(-f)x = f(-x)\ (x \in E)$. If $n \equiv 2 \bmod 4$ then $\alpha(f) \neq \alpha(-f)$, as shown in §6 of [11]. Hence and from the corresponding result for homotopy-symmetry we obtain

Theorem 2.6. Let $n \equiv 2 \bmod 4$. If E admits an A-structure (homotopy-symmetry) of one type then E admits an A-structure (homotopy-symmetry) of the other type.

3. Manifolds

Let M be a compact Riemannian manifold. Let X be a field of unit tangent vectors on M with finite singularities (i.e. X is undefined on a finite set). Homotopies between two such fields are defined in the obvious way. We describe a homotopy X_t as a homotopy-symmetry of X if $X_o = X$ and $X_1 = -X$. Note that a field (X,Y) of orthonormal pairs, with finite singularities, determines a homotopy-symmetry X_t where (3.1) $X_t = X \cos \pi t + Y \sin \pi t \ (t \in I)$.

We refer to the survey article of Thomas [22] for general information about index theory. In particular we recall the celebrated result of Hopf which establishes that

$$\text{Index}(X) = \chi(M), \tag{3.2}$$

where $\chi(M)$ denotes the Euler-Poincaré characteristic. Various further results are known concerning the index of a field of orthonormal pairs, and more generally.

The index of a homotopy-symmetry X_t can be defined as follows. Around each of the singularities X_t defines, in an obvious way, a homotopy-symmetry of a (non-singular) field of unit tangent vectors to S^{n-1}, where $n = \dim M$. Each of these determines a map of S^{n-1} into V''_n and hence an element of $\pi_{n-1}(V''_n)$. We define $\text{Index}(X_t)$ to be the sum of these elements, with orientations taken into account

Note that if X_t is a homotopy-symmetry of X then

(3.3) $$\mathrm{Index}(X) = p''_*\mathrm{Index}(X_t).$$

Also note that if X_t is given by a field (X,Y) of orthonormal pairs, as in (3.1), then

(3.4) $$\mathrm{Index}(X_t) = h_*\mathrm{Index}(X,Y).$$

When n is odd and $n \geq 5$ we have that $\pi_{n-1}(V''_n) \approx Z_2$. When n is even and $n \geq 4$ we have that $\pi_{n-1}(V''_n) \approx Z \oplus Z_2$, and write

$$\mathrm{Index}(X_t) = (Z\text{-}\mathrm{Index}(X_t),\ Z_2\text{-}\mathrm{Index}(X_t)),$$

where $Z\text{-}\mathrm{Index}(X_t) = \mathrm{Index}(X)$, by (3.3).

Each of the standard theorems on the index of a field of orthonormal pairs has an analogue in the homotopy-symmetry case. Let $k(M) \in Z_2$ denote the real Kervaire characteristic of M, which is defined when n is odd. Let $\sigma(M) \in Z$ denote the signature of M, which is defined when $n \equiv 0 \bmod 4$ and which has the same parity as $\chi(M)$. The four cases are as follows.

<u>Theorem</u> 3.5. <u>Let</u> $n \equiv 3 \bmod 4$ <u>and</u> $n \geq 7$. <u>Then</u> $\mathrm{Index}(X_t) = 0$.

<u>Theorem</u> 3.6. <u>Let</u> $n \equiv 1 \bmod 4$ <u>and</u> $n \geq 5$. <u>Then</u> $\mathrm{Index}(X_t) = k(M)$.

<u>Theorem</u> 3.7. <u>Let</u> $n \equiv 2 \bmod 4$ <u>and</u> $n \geq 6$. <u>Then</u> $Z_2\text{-}\mathrm{Index}(X_t) = 0$.

<u>Theorem</u> 3.8. <u>Let</u> $n = 4k$, <u>where</u> $k \geq 1$. <u>Then</u> $Z_2\text{-}\mathrm{Index}(X_t) = (\tfrac{1}{2}(\sigma(M) - (-1)^k\chi(M)) \bmod 2$.

Perhaps the best method of proving these results is to adapt the Postnikov-theoretical versions of the corresponding results for the index of a field of orthonormal pairs. The modifications necessary are purely nominal except there is no necessity to treat the case $n = 4$ separately. An alternative method is to show that every homotopy-symmetry can be deformed into one which is obtainable, as in (3.1), from a field of orthonormal pairs, and then apply (3.4). However, this is open to aesthetic objections in case $n = 4$. It may well be that the analytical techniques of Atiyah [2] can be developed so as to give an independent proof but I have not looked into this possibility.

Now let E be the $(n-1)$-sphere bundle of unit tangent vectors to M. If E admits an A-structure (homotopy-symmetry) we simply say that M admits an A-structure (homotopy-symmetry). Suppose that we have a homotopy-symmetry $g_t : E \to E$, of type $\beta(g_t) \in \pi_{n-1}(V''_n)$. If X is a field of unit tangent vectors, with finite singularities, the $g_t X$ is a homotopy-symmetry of X, such that

(3.9) $$\operatorname{Index}(g_t X) = \beta(g_t) \operatorname{Index}(X).$$

Hence and from (3.8) we obtain

<u>Corollary</u> 3.10. <u>Let</u> $n \equiv 0 \bmod 4$. <u>If</u> M <u>admits a homotopy-symmetry and</u> $\chi(M)$ <u>is even then</u> $\sigma(M) \equiv \chi(M) \bmod 4$.

This shows, for example, that the sphere S^n is not homotopy-symmetric when $n \equiv 0 \bmod 4$. In fact one of the main purposes of [9] was to prove, along these lines, that neither S^4 nor S^8 admits an A-structure. However, it can be shown by direct construction, as in

§4, that the only homotopy-symmetric spheres are S^2 and S^6. Another application of (3.10) is to show that a quaternionic projective k-space is not homotopy-symmetric for odd values of k. In fact by combining the main theorem of [10] with the argument given by Massey [18] we can extend this result to include even values of k as well.

4. Homotopy-symmetric spheres

<u>Theorem</u> 4.1. <u>The sphere</u> S^n <u>admits a homotopy-symmetry if and only if</u> $n = 2$ <u>or</u> 6.

If $n = 2$ or 6 then S^n admits orthogonal A-structure, and so the first part of (4.1) follows at once. To prove the second part we show by a direct construction, that a homotopy-symmetry on S^n determines an H-structure on S^{n+1}. Points of S^{n+1} are represented in the form

$$(e \cos \theta + u \sin \theta) \quad (u \in S^n, t \in I)$$

in the usual way, where e is the "north pole". Points of the tangent sphere-bundle of S^n are represented by orthonormal pairs (u,v), where $u,v \in S^n$, so that the fibration is given by the first component. From the second component of a homotopy-symmetry we obtain a deformation $h_t: E \to S^n$, where $-1 \leq 1 \leq 1$, such that $h_t(u,v) = tv (t = \pm 1)$. This deformation determines an H-structure

$$m: S^{n+1} \times S^{n+1} \to S^{n+1}$$

as follows. If $x,y \in S^n$ and $\theta, \phi \in [0,\pi]$ we define

$$m(e \cos \theta + x \sin \theta,\ e \cos \phi + x \sin \phi) =$$

$$(e \cos\theta + x \sin\theta)\cos\phi + (-e \sin\theta + x \cos\theta)\sin\phi \cos\psi$$
$$+ \sin\psi \sin\phi \, h_{\cos\theta}(x, \frac{y - x\cos\psi}{\sin\psi}),$$

where $\psi \in [0,\pi]$ denotes the angle between x and y. It is easy to check that m is a well-defined continuous function which constitutes an H-structure on S^{n+1}. Since $S^{n+1} (n > 0)$ is not an H-space unless $n = 2$ or 6 this compleles the proof of (4.1). This proof is taken from [16]; an alternative proof is given in [15].

REFERENCES

[1] Adams, J.F., Vector fields on spheres, Annals of Math. 75(1962), 603-662.

[2] Atiyah, M.F., Vector fields on manifolds, Arbeitsgemeinschaft für Forschung des Landes Nordrhein-Westfalen, Heft 200.

[3] Atiyah, M.F. and Dupont, J.L., Vector fields with finite singularities, Acta Mathematica 128(1972), 1-40.

[4] Becker, J.C., On the existence of A_k-structures on stable vector bundles, Topology 9(1970), 367-384.

[5] Bott, R., The stable homotopy of the classical groups, Annals of Math. 70(1959), 313-337.

[6] Dold, A., Uber fasernweise Homotopieaquivalenz von Faserraumen, Math. Zeitschrift 62(1955), 111-136.

[7] Hirzebruch, F. and Hopf, H., Felder von flächenelementen in 4-dimensionalen mannigkeiten, Math. Ann. 136(1958), 156-172.

[8] Hopf, H., Vectorfelder in n-dimensionalen mannigfaltighkeiten, Math. Ann. 96(1927), 225-260.

[9] Hopf, H., Zur topologie de komplexen mannigfaltigkeiten, Studies and Essays presented to R. Courant, Interscience 1948.

[10] James, I.M., Bundles with special structure I, Annals of Math. 89(1969), 359-360.

[11] James, I.M., On fibre bundles and their homotopy groups, J. of Math. Kyoto Univ. 9(1969), 5-24.

[12] James, I.M., On the Bott suspension, J. of Math. Kyoto Univ. 9 (1969), 161-188.

[13] James, I.M., On sphere-bundles I, Bull. Amer. Math. Soc. 75(1969)

617-621.

[14] James, I.M., On sphere-bundles II, Bull. London Math. Soc. 1 (1969), 323-328.

[15] James, I.M., On sphere-bundles with certain properties, Quart. J. Math. Oxford (2), 22(1971), 353-370.

[16] James, I.M., On the homotopy-symmetry of sphere-bundles, Proc. Camb. Phil. Soc. 69(1971), 291-294.

[17] Kervaire, M.A., Relative characteristic classes, American J. of Math. 79(1957), 517-558.

[18] Massey, W.S., Non-existence of almost-complex structures on quaternionic projective spaces, Pacific J. Math. 12 (1962), 1379-1384.

[19] Steenrod, N.E., Topology of fibre bundles, Princeton 1950.

[20] Thomas, E., On tensor products of n-plane bundles, Archiv de Mathematik 10(1959), 174-179.

[21] Thomas, E., The index of a tangent 2-field, Comment. Math. Helv. 42(1967), 86-110.

[22] Thomas, E., Vector fields on manifolds, Bull. Amer. Math. Soc. 75(1969), 643-683.

cture Notes in Mathematics

rehensive leaflet on request

: A. Fröhlich, Formal Groups. IV, 140 pages. 1968. DM 16,–

5: G. Lumer, Algèbres de fonctions et espaces de Hardy. VI, es. 1968. DM 16,–

6: R. G. Swan, Algebraic K-Theory. IV, 262 pages. 1968.
–

7: P.-A. Meyer, Processus de Markov: la frontière de Martin. IV, ges. 1968. DM 16,–

8: H. Herrlich, Topologische Reflexionen und Coreflexionen. XVI, iten. 1968. DM 16,–

9: A. Grothendieck, Catégories Cofibrées Additives et Complexe gent Relatif. IV, 167 pages. 1968. DM 16,–

0: Seminar on Triples and Categorical Homology Theory. Edited Eckmann. IV, 398 pages. 1969. DM 20,–

: J.-P. Eckmann et M. Guenin, Méthodes Algébriques en Méca- Statistique. VI, 131 pages. 1969. DM 16,–

2: J. Wloka, Grundräume und verallgemeinerte Funktionen. VIII, iten. 1969. DM 16,–

3: O. Zariski, An Introduction to the Theory of Algebraic Surfaces. d Printing. IV, 100 pages. 1972. DM 16,–

4: H. Lüneburg, Transitive Erweiterungen endlicher Permutations- en. IV, 119 Seiten. 1969. DM 16,–

5: P. Cartier et D. Foata, Problèmes combinatoires de commu- et réarrangements. IV, 88 pages. 1969. DM 16,–

6: Category Theory, Homology Theory and their Applications I. by P. Hilton. VI, 216 pages. 1969. DM 16,–

7: M. Tierney, Categorical Constructions in Stable Homotopy . IV, 65 pages. 1969. DM 16,–

8: Séminaire de Probabilités III. IV, 229 pages. 1969.
–

9: Probability and Information Theory. Edited by M. Behara, keberg and J. Wolfowitz. IV, 256 pages. 1969. DM 18,–

0: N. P. Bhatia and O. Hajek, Local Semi-Dynamical Systems pages. 1969. DM 16,–

: N. N. Janenko, Die Zwischenschrittmethode zur Lösung mehr- sionaler Probleme der mathematischen Physik. VIII, 194 Seiten. DM 16,80

2: Category Theory, Homology Theory and their Applications II by P. Hilton. V, 308 pages. 1969. DM 20,–

3: K. R. Parthasarathy, Multipliers on Locally Compact Groups. pages. 1969. DM 16,–

4: M. Machover and J. Hirschfeld, Lectures on Non-Standard sis. VI, 79 pages. 1969. DM 16,–

5: A. S. Troelstra, Principles of Intuitionism. II, 111 pages. 1969. ,–

6: H.-B. Brinkmann und D. Puppe, Abelsche und exakte Kate- , Korrespondenzen. V, 141 Seiten. 1969. DM 16,–

7: S. O. Chase and M. E. Sweedler, Hopf Algebras and Galois II, 133 pages. 1969. DM 16,–

8: M. Heins, Hardy Classes on Riemann Surfaces. III, 106 pages. DM 16,–

9: Category Theory, Homology Theory and their Applications III. by P. Hilton. IV, 489 pages. 1969. DM 24,–

0: M. Artin and B. Mazur, Etale Homotopy. II, 196 Seiten. 1969 ,–

1: G. P. Szegö et G. Treccani, Semigruppi di Trasformazioni oche. VI, 177 pages. 1969. DM 16,–

2: F. Stummel, Rand- und Eigenwertaufgaben in Sobolewschen n. VIII, 386 Seiten. 1969. DM 20,–

3: Lectures in Modern Analysis and Applications I. Edited by aam. VII, 162 pages. 1969. DM 16,–

4: G. H. Pimbley, Jr., Eigenfunction Branches of Nonlinear ors and their Bifurcations. II, 128 pages. 1969. DM 16,–

5: R. Larsen, The Multiplier Problem. VII, 284 pages. 1969.
–

6: Reports of the Midwest Category Seminar III. Edited by S. Mac , 247 pages. 1969. DM 16,–

7: A. Peyerimhoff, Lectures on Summability. III, 111 pages. DM 16,–

8: Algebraic K-Theory and its Geometric Applications. Edited F. Moss and C. B. Thomas. IV, 86 pages. 1969. DM 16,–

9: Conference on the Numerical Solution of Differential Equa- Edited by J. Ll. Morris. VI, 275 pages. 1969. DM 18,–

0: The Many Facets of Graph Theory. Edited by G. Chartrand F. Kapoor. VIII, 290 pages. 1969. DM 18,–

Vol. 111: K. H. Mayer, Relationen zwischen charakteristischen Zahlen. III, 99 Seiten. 1969. DM 16,–

Vol. 112: Colloquium on Methods of Optimization. Edited by N. N. Moiseev. IV, 293 pages. 1970. DM 18,–

Vol. 113: R. Wille, Kongruenzklassengeometrien. III, 99 Seiten. 1970. DM 16,–

Vol. 114: H. Jacquet and R. P. Langlands, Automorphic Forms on GL (2). VII, 548 pages. 1970. DM 24,–

Vol. 115: K. H. Roggenkamp and V. Huber-Dyson, Lattices over Orders I. XIX, 290 pages. 1970. DM 18,–

Vol. 116: Séminaire Pierre Lelong (Analyse) Année 1969. IV, 195 pages. 1970. DM 16,–

Vol. 117: Y. Meyer, Nombres de Pisot, Nombres de Salem et Analyse Harmonique. 63 pages. 1970. DM 16,–

Vol. 118: Proceedings of the 15th Scandinavian Congress, Oslo 1968. Edited by K. E. Aubert and W. Ljunggren. IV, 162 pages. 1970. DM 16,–

Vol. 119: M. Raynaud, Faisceaux amples sur les schémas en groupes et les espaces homogènes. III, 219 pages. 1970. DM 16,–

Vol. 120: D. Siefkes, Büchi's Monadic Second Order Successor Arithmetic. XII, 130 Seiten. 1970. DM 16,–

Vol. 121: H. S. Bear, Lectures on Gleason Parts. III, 47 pages. 1970. DM 16,–

Vol. 122: H. Zieschang, E. Vogt und H.-D. Coldewey, Flächen und ebene diskontinuierliche Gruppen. VIII, 203 Seiten. 1970. DM 16,–

Vol. 123: A. V. Jategaonkar, Left Principal Ideal Rings. VI, 145 pages. 1970. DM 16,–

Vol. 124: Séminare de Probabilités IV. Edited by P. A. Meyer. IV, 282 pages. 1970. DM 20,–

Vol. 125: Symposium on Automatic Demonstration. V, 310 pages. 1970. DM 20,–

Vol. 126: P. Schapira, Théorie des Hyperfonctions. XI, 157 pages. 1970. DM 16,–

Vol. 127: I. Stewart, Lie Algebras. IV, 97 pages. 1970. DM 16,–

Vol. 128: M. Takesaki, Tomita's Theory of Modular Hilbert Algebras and its Applications. II, 123 pages. 1970. DM 16,–

Vol. 129: K. H. Hofmann, The Duality of Compact Semigroups and C*- Bigebras. XII, 142 pages. 1970. DM 16,–

Vol. 130: F. Lorenz, Quadratische Formen über Körpern. II, 77 Seiten. 1970. DM 16,–

Vol. 131: A. Borel et al., Seminar on Algebraic Groups and Related Finite Groups. VII, 321 pages. 1970. DM 22,–

Vol. 132: Symposium on Optimization. III, 348 pages. 1970. DM 22,–

Vol. 133: F. Topsøe, Topology and Measure. XIV, 79 pages. 1970. DM 16,–

Vol. 134: L. Smith, Lectures on the Eilenberg-Moore Spectral Sequence. VII, 142 pages. 1970. DM 16,–

Vol. 135: W. Stoll, Value Distribution of Holomorphic Maps into Compact Complex Manifolds. II, 267 pages. 1970. DM 18,–

Vol. 136: M. Karoubi et al., Séminaire Heidelberg-Saarbrücken-Strasbuorg sur la K-Théorie. IV, 264 pages. 1970. DM 18,–

Vol. 137: Reports of the Midwest Category Seminar IV. Edited by S. MacLane. III, 139 pages. 1970. DM 16,–

Vol. 138: D. Foata et M. Schützenberger, Théorie Géométrique des Polynômes Eulériens. V, 94 pages. 1970. DM 16,–

Vol. 139: A. Badrikian, Séminaire sur les Fonctions Aléatoires Linéaires et les Mesures Cylindriques. VII, 221 pages. 1970. DM 18,–

Vol. 140: Lectures in Modern Analysis and Applications II. Edited by C. T. Taam. VI, 119 pages. 1970. DM 16,–

Vol. 141: G. Jameson, Ordered Linear Spaces. XV, 194 pages. 1970. DM 16,–

Vol. 142: K. W. Roggenkamp, Lattices over Orders II. V, 388 pages. 1970. DM 22,–

Vol. 143: K. W. Gruenberg, Cohomological Topics in Group Theory. XIV, 275 pages. 1970. DM 20,–

Vol. 144: Seminar on Differential Equations and Dynamical Systems, II. Edited by J. A. Yorke. VIII, 268 pages. 1970. DM 20,–

Vol. 145: E. J. Dubuc, Kan Extensions in Enriched Category Theory. XVI, 173 pages. 1970. DM 16,–

Please turn over

Vol. 146: A. B. Altman and S. Kleiman, Introduction to Grothendieck Duality Theory. II, 192 pages. 1970. DM 18,–

Vol. 147: D. E. Dobbs, Cech Cohomological Dimensions for Commutative Rings. VI, 176 pages. 1970. DM 16,–

Vol. 148: R. Azencott, Espaces de Poisson des Groupes Localement Compacts. IX, 141 pages. 1970. DM 16,–

Vol. 149: R. G. Swan and E. G. Evans, K-Theory of Finite Groups and Orders. IV, 237 pages. 1970. DM 20,–

Vol. 150: Heyer, Dualität lokalkompakter Gruppen. XIII, 372 Seiten. 1970. DM 20,–

Vol. 151: M. Demazure et A. Grothendieck, Schémas en Groupes I. (SGA 3). XV, 562 pages. 1970. DM 24,–

Vol. 152: M. Demazure et A. Grothendieck, Schémas en Groupes II. (SGA 3). IX, 654 pages. 1970. DM 24,–

Vol. 153: M. Demazure et A. Grothendieck, Schémas en Groupes III. (SGA 3). VIII, 529 pages. 1970. DM 24,–

Vol. 154: A. Lascoux et M. Berger, Variétés Kähleriennes Compactes. VII, 83 pages. 1970. DM 16,–

Vol. 155: Several Complex Variables I, Maryland 1970. Edited by J. Horváth. IV, 214 pages. 1970. DM 18,–

Vol. 156: R. Hartshorne, Ample Subvarieties of Algebraic Varieties. XIV, 256 pages. 1970. DM 20,–

Vol. 157: T. tom Dieck, K. H. Kamps und D. Puppe, Homotopietheorie. VI, 265 Seiten. 1970. DM 20,–

Vol. 158: T. G. Ostrom, Finite Translation Planes. IV. 112 pages. 1970. DM 16,–

Vol. 159: R. Ansorge und R. Hass. Konvergenz von Differenzenverfahren für lineare und nichtlineare Anfangswertaufgaben. VIII, 145 Seiten. 1970. DM 16,–

Vol. 160: L. Sucheston, Constributions to Ergodic Theory and Probability. VII, 277 pages. 1970. DM 20,–

Vol. 161: J. Stasheff, H-Spaces from a Homotopy Point of View. VI, 95 pages. 1970. DM 16,–

Vol. 162: Harish-Chandra and van Dijk, Harmonic Analysis on Reductive p-adic Groups. IV, 125 pages. 1970. DM 16,–

Vol. 163: P. Deligne, Equations Différentielles à Points Singuliers Reguliers. III, 133 pages. 1970. DM 16,–

Vol. 164: J. P. Ferrier, Seminaire sur les Algebres Complètes. II, 69 pages. 1970. DM 16,–

Vol. 165: J. M. Cohen, Stable Homotopy. V, 194 pages. 1970. DM 16,–

Vol. 166: A. J. Silberger, PGL_2 over the p-adics: its Representations, Spherical Functions, and Fourier Analysis. VII, 202 pages. 1970. DM 18,–

Vol. 167: Lavrentiev, Romanov and Vasiliev, Multidimensional Inverse Problems for Differential Equations. V, 59 pages. 1970. DM 16,–

Vol. 168: F. P. Peterson, The Steenrod Algebra and its Applications: A conference to Celebrate N. E. Steenrod's Sixtieth Birthday. VII, 317 pages. 1970. DM 22,–

Vol. 169: M. Raynaud, Anneaux Locaux Henséliens. V, 129 pages. 1970. DM 16,–

Vol. 170: Lectures in Modern Analysis and Applications III. Edited by C. T. Taam. VI, 213 pages. 1970. DM 18,–

Vol. 171: Set-Valued Mappings, Selections and Topological Properties of 2^X. Edited by W. M. Fleischman. X, 110 pages. 1970. DM 16,–

Vol. 172: Y.-T. Siu and G. Trautmann, Gap-Sheaves and Extension of Coherent Analytic Subsheaves. V, 172 pages. 1971. DM 16,–

Vol. 173: J. N. Mordeson and B. Vinograde, Structure of Arbitrary Purely Inseparable Extension Fields. IV, 138 pages. 1970. DM 16,–

Vol. 174: B. Iversen, Linear Determinants with Applications to the Picard Scheme of a Family of Algebraic Curves. VI, 69 pages. 1970. DM 16,–

Vol. 175: M. Brelot, On Topologies and Boundaries in Potential Theory. VI, 176 pages. 1971. DM 18,–

Vol. 176: H. Popp, Fundamentalgruppen algebraischer Mannigfaltigkeiten. IV, 154 Seiten. 1970. DM 16,–

Vol. 177: J. Lambek, Torsion Theories, Additive Semantics and Rings of Quotients. VI, 94 pages. 1971. DM 16,–

Vol. 178: Th. Bröcker und T. tom Dieck, Kobordismentheorie. XVI, 191 Seiten. 1970. DM 18,–

Vol. 179: Seminaire Bourbaki – vol. 1968/69. Exposés 347-363. IV 295 pages. 1971. DM 22,–

Vol. 180: Séminaire Bourbaki – vol. 1969/70. Exposés 364-38 310 pages. 1971. DM 22,–

Vol. 181: F. DeMeyer and E. Ingraham, Separable Algebras Commutative Rings. V, 157 pages. 1971. DM 16.–

Vol. 182: L. D. Baumert. Cyclic Difference Sets. VI, 166 pages. DM 16,–

Vol. 183: Analytic Theory of Differential Equations. Edited by P. F. and A. W. J. Stoddart. VI, 225 pages. 1971. DM 20,–

Vol. 184: Symposium on Several Complex Variables, Park City, 1970. Edited by R. M. Brooks. V, 234 pages. 1971. DM 20,–

Vol. 185: Several Complex Variables II, Maryland 1970. Edit J. Horváth. III, 287 pages. 1971. DM 24,–

Vol. 186: Recent Trends in Graph Theory. Edited by M. Capob J. B. Frechen/M. Krolik. VI, 219 pages. 1971. DM 18.–

Vol. 187: H. S. Shapiro, Topics in Approximation Theory. VIII, 275 p 1971. DM 22,–

Vol. 188: Symposium on Semantics of Algorithmic Languages. E by E. Engeler. VI, 372 pages. 1971. DM 26,–

Vol. 189: A. Weil, Dirichlet Series and Automorphic Forms. V pages. 1971. DM 16,–

Vol. 190: Martingales. A Report on a Meeting at Oberwolfach 17-23, 1970. Edited by H. Dinges. V, 75 pages. 1971. DM 16,–

Vol. 191: Séminaire de Probabilités V. Edited by P. A. Meyer. I pages. 1971. DM 26,–

Vol. 192: Proceedings of Liverpool Singularities – Symposium I. by C. T. C. Wall. V, 319 pages. 1971. DM 24,–

Vol. 193: Symposium on the Theory of Numerical Analysis. by J. Ll. Morris. VI, 152 pages. 1971. DM 16,–

Vol. 194: M. Berger, P. Gauduchon et E. Mazet. Le Spectre Variété Riemannienne. VII, 251 pages. 1971. DM 22,–

Vol. 195: Reports of the Midwest Category Seminar V. Edited b Gray and S. Mac Lane. III, 255 pages. 1971. DM 22,–

Vol. 196: H-spaces – Neuchâtel (Suisse)- Août 1970. Edited Sigrist, V, 156 pages. 1971. DM 16,–

Vol. 197: Manifolds – Amsterdam 1970. Edited by N. H. Kuiper. pages. 1971 DM 20,–

Vol. 198: M. Hervé, Analytic and Plurisubharmonic Functions in and Infinite Dimensional Spaces. VI, 90 pages. 1971. DM 16.–

Vol. 199: Ch. J. Mozzochi, On the Pointwise Convergence of F Series. VII, 87 pages. 1971. DM 16,–

Vol. 200: U. Neri, Singular Integrals. VII, 272 pages. 1971. DM

Vol. 201: J. H. van Lint, Coding Theory. VII, 136 pages. 1971. DM

Vol. 202: J. Benedetto, Harmonic Analysis on Totally Disconn Sets. VIII, 261 pages. 1971. DM 22,–

Vol. 203: D. Knutson, Algebraic Spaces. VI, 261 pages. DM 22,–

Vol. 204: A. Zygmund, Intégrales Singulières. IV, 53 pages. DM 16,–

Vol. 205: Séminaire Pierre Lelong (Analyse) Année 1970. VI, pages. 1971. DM 20,–

Vol. 206: Symposium on Differential Equations and Dynamica tems. Edited by D. Chillingworth. XI, 173 pages. 1971. DM

Vol. 207: L. Bernstein, The Jacobi-Perron Algorithm – Its T and Application. IV, 161 pages. 1971. DM 16,–

Vol. 208: A. Grothendieck and J. P. Murre, The Tame Fundam Group of a Formal Neighbourhood of a Divisor with Normal Cros on a Scheme. VIII, 133 pages. 1971. DM 16,–

Vol. 209: Proceedings of Liverpool Singularities Symposi Edited by C. T. C. Wall. V, 280 pages. 1971. DM 22,–

Vol. 210: M. Eichler, Projective Varieties and Modular Forms. II pages. 1971. DM 16,–

Vol. 211: Théorie des Matroïdes. Edité par C. P. Bruter. III pages. 1971. DM 16,–

Vol. 212: B. Scarpellini, Proof Theory and Intuitionistic Sys VII, 291 pages. 1971. DM 24,–

Vol. 213: H. Hogbe-Nlend, Théorie des Bornologies et Applica V, 168 pages. 1971. DM 18,–

Vol. 214: M. Smorodinsky, Ergodic Theory, Entropy. V, 64 p 1971. DM 16,–